Springer Theses

Recognizing Outstanding Ph.D. Research

For further volumes:
http://www.springer.com/series/8790

Aims and Scope

The series "Springer Theses" brings together a selection of the very best Ph.D. theses from around the world and across the physical sciences. Nominated and endorsed by two recognized specialists, each published volume has been selected for its scientific excellence and the high impact of its contents for the pertinent field of research. For greater accessibility to non-specialists, the published versions include an extended introduction, as well as a foreword by the student's supervisor explaining the special relevance of the work for the field. As a whole, the series will provide a valuable resource both for newcomers to the research fields described, and for other scientists seeking detailed background information on special questions. Finally, it provides an accredited documentation of the valuable contributions made by today's younger generation of scientists.

Theses are accepted into the series by invited nomination only and must fulfill all of the following criteria

- They must be written in good English.
- The topic should fall within the confines of Chemistry, Physics, Earth Sciences, Engineering and related interdisciplinary fields such as Materials, Nanoscience, Chemical Engineering, Complex Systems and Biophysics.
- The work reported in the thesis must represent a significant scientific advance.
- If the thesis includes previously published material, permission to reproduce this must be gained from the respective copyright holder.
- They must have been examined and passed during the 12 months prior to nomination.
- Each thesis should include a foreword by the supervisor outlining the significance of its content.
- The theses should have a clearly defined structure including an introduction accessible to scientists not expert in that particular field.

Ravi Kumar Pujala

Dispersion Stability, Microstructure and Phase Transition of Anisotropic Nanodiscs

Doctoral Thesis accepted by
Jawaharlal Nehru University, New Delhi, India

Author
Dr. Ravi Kumar Pujala
School of Physical Sciences
Jawaharlal Nehru University
New Delhi
India

Supervisor
Prof. Himadri B. Bohidar
School of Physical Sciences
Jawaharlal Nehru University
New Delhi
India

ISSN 2190-5053 ISSN 2190-5061 (electronic)
ISBN 978-3-319-04554-2 ISBN 978-3-319-04555-9 (eBook)
DOI 10.1007/978-3-319-04555-9
Springer Cham Heidelberg New York Dordrecht London

Library of Congress Control Number: 2014939646

© Springer International Publishing Switzerland 2014
This work is subject to copyright. All rights are reserved by the Publisher, whether the whole or part of the material is concerned, specifically the rights of translation, reprinting, reuse of illustrations, recitation, broadcasting, reproduction on microfilms or in any other physical way, and transmission or information storage and retrieval, electronic adaptation, computer software, or by similar or dissimilar methodology now known or hereafter developed. Exempted from this legal reservation are brief excerpts in connection with reviews or scholarly analysis or material supplied specifically for the purpose of being entered and executed on a computer system, for exclusive use by the purchaser of the work. Duplication of this publication or parts thereof is permitted only under the provisions of the Copyright Law of the Publisher's location, in its current version, and permission for use must always be obtained from Springer. Permissions for use may be obtained through RightsLink at the Copyright Clearance Center. Violations are liable to prosecution under the respective Copyright Law.
The use of general descriptive names, registered names, trademarks, service marks, etc. in this publication does not imply, even in the absence of a specific statement, that such names are exempt from the relevant protective laws and regulations and therefore free for general use.
While the advice and information in this book are believed to be true and accurate at the date of publication, neither the authors nor the editors nor the publisher can accept any legal responsibility for any errors or omissions that may be made. The publisher makes no warranty, express or implied, with respect to the material contained herein.

Printed on acid-free paper

Springer is part of Springer Science+Business Media (www.springer.com)

*This thesis is dedicated to my parents
for their endless love, support
and encouragement...*

Supervisor's Foreword

Some of the most remarkable recent advances in colloid science have exploited the specific pairing of clay platelets to create dynamic three-dimensional structures that precisely self-assemble from individual clay particles. Pioneering work in this area has been reported by various research groups in the recent past. Colloidal self-assembly often leads to gel or glass-like organization of internal structure.

Colloidal gels can be thought of as a space filling or percolating network of particles and particulate suspensions, which can form when the system is destabilized. Under appropriate thermodynamic conditions, sufficient particle concentration and provided that the attraction between the particles is strong enough to induce aggregation, a space filling macroscopic structure is formed that effectively traps the solvent molecules. Although not in a thermodynamic equilibrium state, the network can undergo a remarkable kinetic slowdown and the resulting gel attains age-dependent viscoelastic attributes. The important parameter that affects this kinetic slowdown is the range of the attractive potential and its relation to the particle size and density. It is also important whether the contact of two particles results in a permanent bond, or if the particles have the additional freedom of rotating on top of each other. In the first case, the formation of the network is dominated by the diffusion while in the second case there is a slow kinetic evolution driven by the phase separation occurring in the system.

On the other hand, colloidal glasses are concentrated suspensions of microscopic particles in a liquid in which the particles' movements are constrained; they hold some freedom for local Brownian motions, but are unable to diffuse over large lengths. Due to this localization, colloidal glasses at rest are amorphous solids. Nevertheless, they are typically soft solids, deforming elastically under small applied stresses, but yielding and flowing when stressed more strongly.

The objective of this thesis was to develop a general understanding of self-assembly of anisotropically charged platelets through coarse grained model that is simple enough to be experimentally verifiable, but complex enough to capture the structural, thermodynamic, and mechanical properties of the material. By focusing on the basic physics of single platelets, their finite clusters, and the transitions to gel or glass phases, Dr. Ravi Kumar Pujala was able to quantitatively describe many of the underlying physical processes that can be exploited to generate smart functional nanostructures from clay particles. At the same time, detailed analysis of the experimental data helped constrain and improve the theoretical models considerably.

Dr. Ravi Kumar was able to provide detailed information about the pathways explored during a full cycle of colloidal dispersion, gelation, glass formation, and their aging dynamics. These new observations are a step-change improvement over previous attempts to quantify aging dynamics of colloidal self-assembly. The results and conclusions contained in this thesis are thought-provoking and will guide future researchers for a long time to come.

New Delhi, April 2014   Prof. Himadri B. Bohidar

Abstract

This thesis explores the dispersion stability, microstructure and phase transitions involved in the nanoclay system. It describes the recently discovered formation of colloidal gels via two routes: the first is through phase separation and second is by equilibrium gelation and includes the first reported experimental observation of a system with high aspect ratio nanodiscs. The phase behaviour of anisotropic nanodiscs of different aspect ratio in their individual and mixed states in aqueous and hydrophobic media is investigated. Distinct phase separation, equilibrium fluid and equilibrium gel phases are observed in nanoclay dispersions with extensive aging. The work then explores solution behavior, gelation kinetics, aging dynamics and temperature-induced ordering in the individual and mixed states of these discotic colloids. Anisotropic ordering dynamics induced by a water-air interface, waiting time and temperature in these dispersions were studied in great detail along with aggregation behavior of nanoplatelets in hydrophobic environment of alcohol solutions.

Acknowledgments

One of the joys of completion is to look over the journey past and remember all those who have helped and supported me along this long but fulfilling road. I take this opportunity to acknowledge them, to only some of whom it is possible to give particular mention here, and extend my sincere gratitude for helping me make this Ph.D. thesis a possibility.

First and foremost, I would like to express my heartfelt gratitude to my thesis advisor Prof. H. B. Bohidar for the continuous support during my Ph.D. work and research, for his patience, motivation, enthusiasm, and immense knowledge while allowing me the room to work in my own ways. His guidance helped me always at the time of doing research and when writing of this thesis. I could have not imagined having a better advisor and mentor for my Ph.D. work.

Besides my advisor, I would like to thank Prof. Shankar P. Das for his encouragement and new ideas. I would like to acknowledge Prof. R. Ramaswamy, Prof. Sanjay Puri, Prof. Rupamanjari Ghosh, Prof. Deepak Kumar, Prof. Subir K. Sarkar, Prof. S. Patnaik, Prof R. Rajaraman, Prof. Akhilesh Pandey, Dr. Sobhan Sen and Dr. Brijesh Kumar, who taught me during my M.Sc. and Pre-Ph.D. course work and filled confidence in me. Special thanks to Prof. Prasenjit Sen and Dr. Pritam Mukhopadhyay who introduced me to the lab practices during my summer trainings. Special thanks to Dr. Balaji Birajdar for his kind help and encouragement in needy times and I cannot forget his contribution in proofreading.

My sincere gratitude to my seniors—Dr. Pradip, Dr. Santinath, Dr. Nisha, and Dr. Shilpi for their help and the fruitful discussions, colleagues—Dr. Kamla, Shilpa, Najmul, Nidhi and Jyotsana, who have been active and provided cheerful moments and interesting discussions which helped me think better.

I thank all the friends in SPS. You are there in my memory always. I want to thank everyone who has in some way helped me, encouraged me or simply made my research project a truly learning and enriching experience. This includes Council of Scientific and Industrial Research (CSIR), India for financial support, JNU for travel grant to present my work in Germany, staff of SPS and Ad-block for helping me with administrative issues, all AIRF technicians for helping out in obtaining fruitful results.

My heartfelt thanks go to my dear friends Priyanka Nial, Manasa Kandula, Ludhiya and Sowmya, and Srikala akka. I especially acknowledge Delhi Forerunners Community: Abraham and Sarah Nial, Robert and Happy, and Sarah Grace for their love, prayers and support, treated me like a real brother and always made me felt at home.

Above all I would like to acknowledge the tremendous sacrifices that my hardworking parents, Shri. Sivannarayana Pujala (late) and Smt. Obulamma, though they are illiterates, ensured that I had an excellent education. For this and much more, I am forever in their debt. It is to them that I dedicate this thesis. Lots of love for my brother Brahmam, for his guidance, motivation, support all the way in my studies. My sister Pushpa has been my best friend and I love her dearly for all her advice and support. I thank all the support and love from—Shri. Korneli, Yona, Annamani, Nehemya, Supriya, Sunil, Shri and Smt. Koteswararao, Sravani, and Deevena.

Thank you Lord, for always being there for me.

Dr. Ravi Kumar Pujala

Contents

1	**Introduction**		1
	1.1 Soft Matter		1
	1.2 Colloids		2
	1.3 Colloids: Why are Physicists Concerned?		3
	1.4 Interactions in Colloidal System		4
		1.4.1 van der Waals Forces	4
		1.4.2 Double Layer Interactions (Repulsive Potential)	5
		1.4.3 Interaction Energy Between Clay Particles	6
	1.5 Clays as Colloidal Systems		8
	1.6 Description of Nanoclays		8
		1.6.1 Laponite	8
		1.6.2 Montmorillonite	10
	1.7 Objective and Scope of Thesis		11
	1.8 Aging and Dynamic Arrest in Colloidal Systems		11
	1.9 Definition of Various Non-ergodic States		13
		1.9.1 Colloidal Gel	13
		1.9.2 Colloidal Glass	13
		1.9.3 Comparison Between Gel and Glass	13
	1.10 Structure of the Thesis		14
	References		14
2	**Materials and Characterization Techniques**		17
	2.1 Materials Used		17
		2.1.1 Nanoclays	17
		2.1.2 Solvents	17
	2.2 Characterization Techniques		18
		2.2.1 Laser Light Scattering	18
		2.2.2 Rheology	22
		2.2.3 Viscometry	27
		2.2.4 Tensiometry	27
		2.2.5 Electrophoresis	29
		2.2.6 Raman Spectroscopy	32

		2.2.7	Transmission Electron Microscopy	33
		2.2.8	X-Ray Diffraction	34
	References			36

3 Phase Diagram of Aging Laponite Dispersions 37
3.1 Introduction 37
3.2 Sample Preparation 38
3.3 Results and Discussion 38
    3.3.1 Hydration of Laponite 38
    3.3.2 Growth of Structures 42
    3.3.3 Viscoelastic Behaviour 43
    3.3.4 Dispersion Homogeneity at $t_w = 0$ 46
    3.3.5 Phase Diagram 48
3.4 Conclusion 49
References 50

4 Anisotropic Ordering in Nanoclay Dispersions Induced by Water–Air Interface 53
4.1 Introduction 53
4.2 Experimental Geometry 54
4.3 Experimental Results 56
    4.3.1 Anisotropic Ordering at the Water–Air Interface 56
    4.3.2 Effect of Water–Hydrophobic Liquid Interface 58
    4.3.3 Effect of Temperature 58
    4.3.4 Relaxation Dynamics 59
4.4 Discussion 62
4.5 Conclusion 63
References 64

5 Phase Diagram of Aging Montmorillonite Dispersions 67
5.1 Introduction 67
5.2 Sample Preparation 68
5.3 Results and Discussion 69
    5.3.1 Time-Dependent Viscosity 69
    5.3.2 Visco-Elasticity 70
    5.3.3 Steady State Viscosity and Yield Stress 70
    5.3.4 Cole–Cole Plots 75
    5.3.5 Light Scattering Experiments 76
    5.3.6 Observation of Phase Separation and Equilibrium Gels and Phase Diagram 78
    5.3.7 Gelation Kinetics in Percolation Formalism 79
5.4 Conclusions 80
References 81

6	**Sol State Behavior and Gelation Kinetics in Mixed Nanoclay Dispersions**		83
	6.1	Introduction	83
	6.2	Sample Preparation	84
	6.3	Result and Discussion	85
		6.3.1 Sol State Behavior	85
		6.3.2 Gel State Properties of Mixed Clay Dispersions	87
		6.3.3 Visco-Elastic Properties	89
		6.3.4 Gelation Kinetics in Percolation Formalism	94
	6.4	Conclusion	99
	References		100
7	**Aging Dynamics in Mixed Nanoclay Dispersions**		103
	7.1	Introduction	103
	7.2	Sample Preparation	104
	7.3	Results and Discussion	104
		7.3.1 Concentration Dependence	104
		7.3.2 Ergodicity Breaking Time	105
		7.3.3 Relaxation Dynamics	107
		7.3.4 Behavior of the System $\phi < \phi_{cutoff}$	111
		7.3.5 Growth of Anisotropy with Aging	111
		7.3.6 Cole–Cole Plot and Sample Heterogeneity	113
	7.4	Dilution Experiment	115
	7.5	Conclusion	115
	References		116
8	**Thermal Ordering in Mixed Nanoclay Dispersions**		119
	8.1	Introduction	119
	8.2	Results and Discussion	119
	8.3	Application of Landau Theory	124
	8.4	Conclusion	127
	References		128
9	**Aggregation and Scaling Behavior of Nanoclays in Alcohol Solutions**		131
	9.1	Introduction	131
	9.2	Sample Preparation	132
	9.3	Results and Discussion	133
	9.4	Conclusion	138
	References		138
10	**Summary**		141
	10.1	Summary of the Main Results	141
	10.2	Open Problems	144

Appendix 147

Curriculum Vitae 151

Chapter 1
Introduction

Abstract This chapter introduces the basic elements of the soft matter and colloids, gives the motivation, formulates the importance of present work and finally presents the outline of the thesis.

1.1 Soft Matter

Undoubtedly, soft matter science is rapidly growing and this interdisciplinary field of research has equally attracted attention from diverse group of scientists ranging from chemists, physicists to engineers. What is "Soft Matter"? In the words of Noble Laureate, P.G. de Gennes *"What do we mean by soft matter? Americans prefer to call it "complex fluids". This is a rather ugly name, which tends to discourage the young students. But it does indeed bring in two of the major features:* (1) *complexity and* (2) *flexibility*" [3]. In the most general terms, soft condensed matter is a matter that is "*soft*" at room temperature. The term "*soft*" is referred to substances whose molecules can be moved to significant distances by the application of relatively weak forces. Moreover, soft matter is easily deformed by electromagnetic fields, thermal fluctuations and external stresses. The field of soft matter science consists of polymers, colloids, liquid crystals and biological materials. One picture is worth a thousand words, so the entire field of soft matter science is summarized in Fig. 1.1.

Typically the interactions involved in soft matter science can be divided into two main categories: simple and structural. Simple interactions: dipole–dipole interactions, ionic interactions, steric interactions and hydrogen bonding. Structural interaction: excluded volume interaction which is responsible for local order and space filling. Nevertheless, the crossover from simple to structural interactions has to do with the connection between microscopic and mesoscopic scales [10]. My research focuses on colloids, which exhibit numerous interesting and complex features.

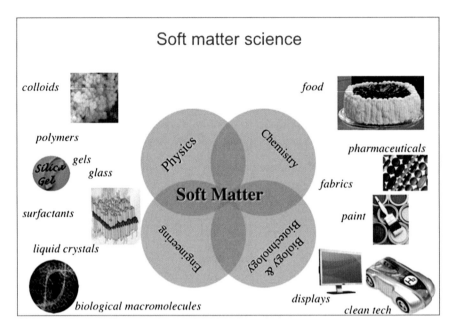

Fig. 1.1 Display of significance of soft matter in various fields of science and technology. (From the lecture "Soft Matter Physics HT08" by Aleksandar Matic)

1.2 Colloids

Colloidal systems of gold particles were already known many centuries ago, and their nature, being "extremely finely divided gold in fluid", was realized as early as 1774 by Juncher and Macquer. Thomas Graham coined the term "colloid" (which means "glue" in Greek) in 1861 to describe Selmi's "pseudosolutions" [6]. This term punctuates their lack of crystallinity and low rate of diffusion. He derived that the low rate of diffusion implied that the particles were fairly large enough than the solvent molecules—in modern terms it is at least 1 nm in size. But then, the absence of particle sedimentation implied an upper size limit of 1 μm. Till date Graham's definition of the range of particle sizes that characterize the colloidal domain is still widely employed. Thus the colloidal systems are solutions of "*large molecules*". The large molecules are *Brownian* or *colloidal particles*. These particle sizes should be large compared to the solvent molecules, but still small enough to exhibit thermal motion in a fluid background solvent. Colloidal solutions are most commonly referred to as *dispersions* or *suspensions* or since here materials is "dispersed" or "suspended" in a liquid phase.

Nowadays colloid science is examining the systems involving particle sizes defined earlier of wide range of systems of scientific and technological importance. Some examples: paints, ceramics, soils, agricultural sprays, cosmetics, biological cells, detergents and many food formulations. Nearly almost all experimental techniques and theoretical procedures of modern physics and modern

1.3 Colloids: Why are Physicists Concerned?

Colloids are dispersions of large molecules (mesoscopic particles) in a fluid background solvent. Here my focus is on the colloidal systems comprising solid, fine particles dispersed in a liquid medium. For a physicist, colloids provide a very good platform for investigating statistical physics, for several main reasons.

1. They are large enough to interact with visible light.
2. They are small enough that their thermal energy $k_B T$ drives their dynamics.
3. Their interactions can be incisively controlled.

To realize the first condition, let us consider a spherical colloidal particle (density $\rho_p = 2.86$ g/ml and diameter $d = 2a = 25$ nm) able to freely diffuse in water (density $\rho_w = 1$ g/ml). The gravitational energy U_g required to lift up the particle its own size is:

$$U_g = m.g.\Delta h = \frac{4}{3}\pi a^3 (\rho_p - \rho_w).g.2a = 6 \times 10^{-29} \text{ J}$$

On the other hand, the thermal energy $k_B T$ at room temperature ($T = 298$ K) is 4.1×10^{-21} J. Evidently $k_B T \gg U_g$, so the particle with relatively small influence of gravity will naturally diffuse throughout the solution on short time scales up to hours or days or months or years. This is in contrast to the granular limit, where gravity essentially immobilizes the particles, and thermal motion is considered irrelevant.

Over the past few years, colloid science has experienced a drastic change, translating itself from little more than accumulation of qualitative observations and ideas of the complex macroscopic behavior of systems into a discipline with well established firm theoretical base. The colloid science now can boast set of conceptions which can provide a better understanding of the many unusual, vague and interesting behavioral patterns displayed by colloidal systems.

To sum up, colloidal dispersions qualify a category of materials that are very common in everyday life. Colloids in industrial products: paints, glues, polishes, lubricants, food products and pharmaceuticals; biological colloids: viruses, bacteria, blood and proteins. Colloidal dispersions are also important as model systems. Their characteristic length- and time-scales are such that they allow for direct experimental observations using advanced microscopy techniques down to single-particle resolution and experimental studies in the light scattering regime with fine time resolution. Since the first experiments of Perrin at the beginning of last century [8], colloids have been used as model systems to study fundamental problems of statistical physics, like crystallization, fluid–fluid phase separation, nucleation, and the wetting of solid substrates. Moreover, the properties of colloidal suspensions can often be changed in such a way that both the strength and the range of the interactions can be manipulated independently, giving rise to complex and fascinating phase behavior, with no similitude in the atomic world [20].

1.4 Interactions in Colloidal System

Derjaguin, Landau, Verwey and Overbeek theory (DLVO theory) proposes that the stability of a particle in solution is dependent upon its total potential energy function V_T. This theory recognizes that V_T is the balance of several competing contributions:

$$V_T = V_A + V_R + V_S \tag{1.1}$$

where V_S is the potential energy due to the solvent, usually it only makes a marginal contribution to the total potential energy over the last few nanometers of separation. Very much important is the balance between V_A and V_R, these are the attractive and repulsive contributions to stabilize the particles.

1.4.1 van der Waals Forces

The attractive forces between neutral and chemically saturated molecules were postulated by van der Waals to explain non-ideal gas behavior. The intermolecular attraction may act between two molecules with permanent dipole mutually oriented towards each other. It may also act between dipolar molecules and neutral molecules in such a way that a dipole is induced in neutral molecules, and then an attraction is generated. Finally intermolecular interactions arise between non polar molecules, where polarization of one molecule is induced by fluctuations in the charge in a second molecule, and vice versa. This last case is known as dispersion force, and it was explained by London in 1930. Apart from highly polar materials, London dispersion force is responsible for all of the van der Waals attraction that acts in a system and it is very short-ranged, since it has an inverse correspondence with the sixth power of the intermolecular distance. For an assembly of molecules the dispersion forces are, to a first approximation, additive in such a way that interaction energy between two particles can be calculated by summing the attractions between all interparticle molecule pairs. Colloidal particles are large assemblies of atoms and hence the result of the sum predicts that the London interaction energy between a group of molecules decays much less rapidly than that between individual molecules. For example, the London dispersion interaction between two spherical particles of radii a_1 and a_2, separated in vacuum by a distance d, was first approximated by Hamaker, [7] and is given as follows:

$$V_A = -\frac{A}{12}\left[\frac{y}{x^2+xy+x} + \frac{y}{x^2+xy+x+y} + 2\ln\left(\frac{x^2+xy+x}{x^2+xy+x+y}\right)\right], \tag{1.2}$$

1.4 Interactions in Colloidal System

where $x = d/(a_1 + a_2)$ and $y = a_1/a_2$.

For spheres with the same size (i.e. $a_1 = a_2 = a$), we have $x = d/2a$, and then Eq. 1.2 becomes

$$V_A = -\frac{A}{12}\left[\frac{1}{x(x+2)} + \frac{1}{(x+1)^2} + 2\ln\left(\frac{x(x+2)}{(x+1)^2}\right)\right] \quad (1.3)$$

As we can see in Eq. 1.3 the attractive potential energy is directly proportional to a particle radius a, to a material constant (called Hamaker constant) A, and is inversely proportional to the separation distance d. Hamaker constant A is given by the expression

$$A = A_{11} + A_{22} - 2A_{12} = \left(A_{11}^{1/2} - A_{22}^{1/2}\right)^2 \quad (1.4)$$

where A_{11} is the Hamaker constant of the particles and A_{22} that of the medium and A_{12} is the Hamaker constant between particle and medium. The Hamaker constant of any material i is related to the London dispersion constant β_{ii} and the number of atoms or molecules per unit volume, q_i, by the expression,

$$A_{ii} = \pi^2 q_i^2 \beta_{ii} \quad (1.5)$$

β_{ii} is related to the polarizability of the atoms or molecules and it has a value in the range 10^{-78}–10^{-76} J m^6 [25].

In the limit of close-approach, the spheres are sufficiently large compared to the distance between them, i.e. ($d \ll 2a$), so that Eq. 1.3 reduces to a simple form:

$$V_A = -\frac{Aa}{12d} \quad (1.6)$$

It means that if the particle approaches very close to each other it will lead to particle coagulation. In fact, the strong Born repulsions between electrons make the value of the van der Waals potential finite at contact, but still many times larger than the thermal energy k_BT, thus leading to an irreversible aggregation. Overlap between colloidal particles are forbidden by Born repulsion will allow the system to be usually modeled by a hard sphere potential for separations $r < 2a$, where r is the center-to-center distance.

1.4.2 Double Layer Interactions (Repulsive Potential)

A charged colloid is surrounded by a solution with inhomogeneous distribution of ions. Co-ions (with same charge as colloids) are depleted from colloidal surface and counterions (with oppositely charge) adsorb at the surface. Far from colloidal surface, concentrations of the two ions attain a constant averaged value. The inhomogeneous layer is termed as double layer and its width depends on ion concentration; adding more ions screens the charges on the colloidal surface. When two double layers overlap, a repulsive pair potential develops which leads to repulsive pressure. Dispersed like-charged colloids hence repel each other upon approach due to screened-Coulomb or double layer repulsion. The interparticle separation dependence of double layer repulsion is approximately exponential [9, 30]

$$V_R = B \frac{R}{l_B} \exp(-\kappa r) \tag{1.7}$$

where κ^{-1} is Debye screening length and R is the radius of the particle.

The quantity B can be expressed in terms of the surface charge density σ_c of the interacting colloids [9, 30]

$$\frac{B}{K_B T} = \frac{8 p_c^2}{1 + p_c^2}, \tag{1.7a}$$

where $p_c = 2\pi \kappa^{-1} l_B |\sigma_c/e|$, with p_c the number of elementary charges e on a surface area $2\pi \kappa^{-1} l_B$. The quantity B can also be expressed as a function of the surface potential ψ_0:

$$\frac{B}{K_B T} = 8 \left[\tanh \left(\frac{\psi_0}{4 K_B T} \right) \right]^2 \tag{1.7b}$$

It is also defined as range of screened double layer repulsion which strongly depends on ionic strength. It is given as

$$\kappa^{-1} = \frac{1}{\sqrt{8\pi l_B n_s}} \tag{1.8}$$

where n_s is concentration of salt and l_B is Bjerrum length which is defined as follows

$$l_B = \frac{e^2}{4\pi \varepsilon_0 \varepsilon_r k_B T} \tag{1.9}$$

where e is elementary charge, k_B is the Boltzman constant, T is temperature, ε_r is relative dielectric constant.

According to DLVO theory

Total interaction = sum of attractive + repulsive interactions;
$$V_T(r) = V_A(r) + V_R(r).$$

1.4.3 Interaction Energy Between Clay Particles

The interaction energy between the clay particles was calculated by DLVO theory as explained above. Using these potentials, a potential barrier high enough to prevent coagulation would exist at all the ionic strengths considered. For very low or zero electrolyte concentration the potential barrier is very high enough to prevent the aggregation as shown in Fig. 1.2. The interaction energy between clay particles in shown in Fig. 1.3.

1.4 Interactions in Colloidal System

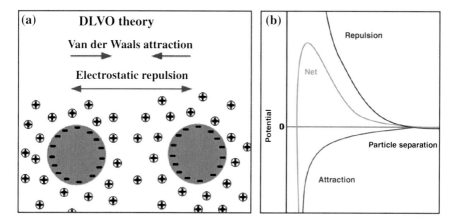

Fig. 1.2 Illustrative DLVO interactions between two charged colloidal particles

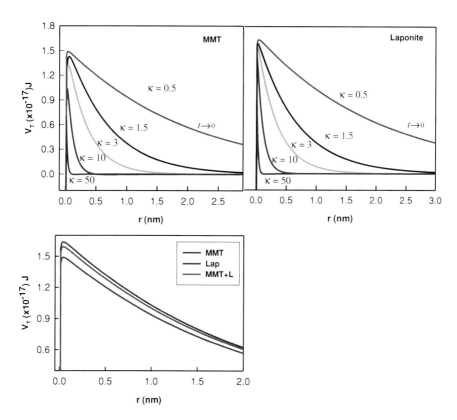

Fig. 1.3 Interaction energy between the clay particles at different ionic strengths as indicated and is shown in the *upper* panel. *Lower* panel shows the comparison of interaction energies of similar particles and mixed particles as indicated

So far colloidal suspensions of spherical particles have been studied in great detail. By contrast, non-spherical colloidal particles have received less attention but the attention is shifting towards them in recent times. Suspensions of anisotropic colloidal particles have unique properties, including their ability to form liquid crystalline phases. Particles having charge and shape anisotropy give numerous interesting phases which are emerging as the new materials in recent times. Inorganic platelets/discs have received more attention from the industrial and scientific point of view.

1.5 Clays as Colloidal Systems

The term "clay" refers to a naturally occurring material composed primarily of fine-grained minerals of diameter generally lower than 2 μm, which behaves as plastic at appropriate water contents and will harden when dried or fired. Even though clay generally contains phyllosilicates, it may contain other materials that give plasticity and harden while dried or fired. Associated phases in clay may include materials that do not impart plasticity and organic matter. Apart from its use in bricks and ceramic, clays are widely used in several industrials applications such as paper making, nanocomposites, chemical filtering, cosmetic material and cement production. Colloidal clays have recently emerged as the complex model systems with variety of phase states, consisting of fluid, gel and glassy states. Theses disordered and amorphous states sometimes interfere with the ordered states called nematic and columnar phases [18, 19]. The role of clay minerals has occupied utmost importance in chemical evolution, the origins of life and these studies are extensively under investigation.

Our focus of this Thesis is on the phase behavior of two clays which belong to phyllosilicate group but have different aspect ratio. One is Laponite, a synthetic clay with diameter of ~25 nm and thickness of ~1 nm (aspect ratio ~25). The other is highly is *Na* Montmorillonite [MMT (*Na* Cloisite)], a naturally occurring anisotropic clay has the diameter of ~250 nm and thickness ~1 nm (aspect ratio ~250). Both the clays exhibit unique properties like the fluid formation, gelation.

Recently, the concept of patchy colloids has emerged for the understanding of the anisotropic potentials and patchy interactions. Clays are also a kind of patchy colloids with very low valence which have the ability to form empty liquids and equilibrium gels. We will give a brief discussion on the two clays including their structures, interactions and phase diagrams.

1.6 Description of Nanoclays

1.6.1 Laponite

Laponite is an entirely synthetic layered silicate developed by Laporte industries between 1965 and 1970 with a structure and composition closely resembling the natural clay mineral hectorite [2]. During the synthesis process, salts of sodium

1.6 Description of Nanoclays

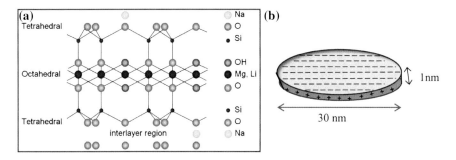

Fig. 1.4 Sketch illustrating the **a** structure of Laponite and **b** the typical dimensions of single Laponite disc

magnesium and lithium are combined with sodium silicate at well controlled rates and temperature. This process results in an amorphous precipitate which is submitted to a high temperature treatment for a partial crystallization. The final product is then filtered, washed, dried and milled, having the appearance of a very fine white powder [12].

Laponite has a layered structure. Each layer is composed of two types of structural sheets: octahedral and tetrahedral. The tetrahedral sheet is composed of silicon-oxygen tetrahedral linked to neighboring tetrahedral by sharing three corners resulting in a hexagonal network. The remaining forth corner of each tetrahedron forms a part from adjacent octahedral sheet. The octahedral sheet is composed of Magnesium in six fold coordination with oxygen from the tetrahedral sheet and with hydroxyl. Figure 1.4a shows the unit cell which depicts six octahedral magnesium ions sandwiched between two layers of four tetrahedral silicon atoms. These groups are balanced by twenty oxygen atoms and four hydroxyl groups. Height of unit cell represents the thickness of laponite crystal. The unit cell is repeated many times in two directions, resulting in the disc shaped appearance of the crystal shown in Fig. 1.4b.

The energy compensation originated by sodium ions is done according to the following molecular formula: $Na^{+0.7}[(Si_8Mg_{5.5}Li_{0.3})O_{20}(OH)_4]^{-0.7}$. The sodium ions are released from the crystal interlayers (see Fig. 1.5b) when dispersed in aqueous solvent, leading to a homogeneous negatively charged faces. A protonation process of the hydroxide groups localized where the crystal structure is terminated produces a positive charge on the rim [26]. Therefore Laponite in aqueous solvent forms a colloidal dispersion of charged disc-like particles with a diameter of ~30 nm and a thickness of ~1 nm with negative and positive charges distributed on the faces and rims respectively [26]. The thickness of each Laponite disc corresponds to the height of the crystal unit cell shown in Fig. 1.4.

An equilibrium is established where sodium ions are held in a diffusive region on both sides of the dispersed laponite crystal. When two particles approach their mutual positive charges repel each other and the dispersion exhibits low viscosity and Newtonian type rheology. Addition of salt to the dispersion of Laponite will reduce the osmotic pressure holding the sodium ions away from the particle

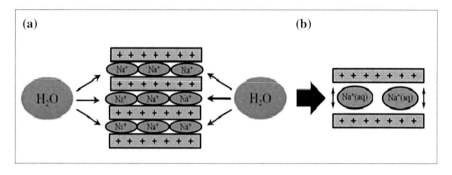

Fig. 1.5 Sketch illustrating **a** the dispersion of Laponite discs in a dry configuration **b** the hydration of the sodium ions

surface. This causes the electrical double layer to contract and allows the weaker positive charge on the edge of the crystal to interact with the negative surface of adjacent crystal. The process may continue to give a "house of cards" structure.

1.6.2 Montmorillonite

Montmorillonite is a very soft phyllosilicate group of minerals that typically form in microscopic crystals, forming a clay. It is named after Montmorillonite in France. Montmorillonite, a member of the smectite family, is a 2:1 clay, meaning that it has 2 tetrahedral sheets sandwiching a central octahedral sheet as shown in Fig. 1.6a. The particles are plate-shaped with an average diameter of approximately one micrometer. Montmorillonite is the main constituent of the volcanic ash weathering product, bentonite. The water content of montmorillonite is variable and it increases greatly in volume when it absorbs water. Chemically it is hydrated sodium calcium aluminium magnesium silicate hydroxide $(Na, Ca)_{0.33}(Al, Mg)_2(Si_4O_{10})(OH)_2 \cdot nH_2O$. Potassium, iron, and other cations are common substitutes; the exact ratio of cations varies with source.

Montmorillonite (MMT) is one of the natural clays, which has high aspect ratio and is a macroscopically swelling, 'active' clay that has the capacity for taking up large amounts of water to form stable gels. The size of the MMT particle is about 250 nm and thickness of the particles is around 1 nm, and the cationic exchange capacity (CEC) is equal to 91 mequiv/100 g of clay. The mechanism for the hydration of cations in Montmorillonite dispersions is alike in Laponite dispersions. These clay particles have the ability to form house of cards structures via face $(-)$/edge $(+)$ (FE) attraction in acidic medium and band-like structures are formed via cation-mediated face $(-)$/face $(-)$ (FF) attraction in alkaline medium. Hence, one has to worry about their full dispersions and pH conditions.

1.7 Objective and Scope of Thesis

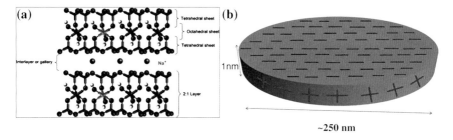

Fig. 1.6 Illustration of **a** structure of Montmorillonite and **b** the typical dimensions of single MMT disc

1.7 Objective and Scope of Thesis

Objective of my thesis is three fold; firstly the study of the phase states involved in the individual nanoclays of Laponite and MMT, secondly the phase behavior of mixed clay system and thirdly the study of phase stability of nanoclays as a function of polarity of solvent.

1.8 Aging and Dynamic Arrest in Colloidal Systems

In condensed matter physics a jammed (glass) state can be obtained by means of temperature quenching. During this process the temperature of a melted material is quickly decreased in such a way that the molecules froze in the previously liquid (disordered) configuration. If a glass former is cooled from its melting temperature to its glass transition temperature T_g, it shows an increase of its relaxation time by 14 decades without a significant change in its structural properties. In soft glassy systems, such as colloids, foams and emulsions, this process is achieved by increasing the concentration of particles or bubbles. Interestingly in colloidal glasses the experiments of Cheng et al., [1] are only able to measure a change in viscosity of 4 orders of magnitude, far less than is seen for molecular glasses. At reasonably high concentrations of the particles the viscosity of the jammed system is so high that due to their crowding or the presence of strong inter particle interactions (strong attractive or repulsive), the mobility of the particles is extremely reduced. These systems are far from the equilibrium configuration and cannot reach the minimum energy configuration even after long wait. Thus, in the process of attaining equilibrium configuration, the properties of the system keep changing with waiting time t_w. Thus, aging is defined as the change in system properties with passage of time. Many jammed system show this behaviour. In a colloidal gel, the correspondence between the relaxation of internal stress and the increase in the gel modulus points strongly to the identification of the internal stress as an indicator for the aging response of the system.

Fig. 1.7 Illustrative picture of the prediction of the mode coupling theory. The dynamic structure factor f(q, t) shows two relaxation times: β for the fast relaxation and α for the slow relaxation. The α mode becomes slower due to the progressive freezing of the intensity fluctuation of the system with waiting time t_w

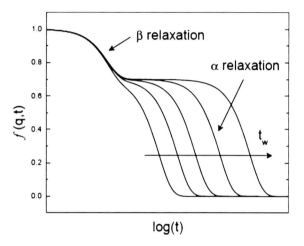

Understanding the slow dynamics in non-equilibrium systems is a long time challenge in physics. Mode coupling theory gives the better understanding of the relaxation dynamics in glasses though there are controversies present in the literature. The evolution of the structural relaxation upon cooling is the theme of the mode-coupling-theory (MCT) of the glass transition [5]. This theory describes the dynamics of a liquid by a set of non-linear equations of motion, which use the equilibrium structure functions as input.

The MCT has attracted much attention due to the interesting predictions regarding the aging and the glass transition of glassy systems in terms of density-density correlation function. The theory identifies that beyond the time scale of microscopic motions, the relaxation of the intermediate scatterings function (density autocorrelation) proceeds in two stages, as shown in Fig. 1.7.

The β or fast relaxation is associated with the rapid motion of the particles inside the cage formed by its neighbors. In other words, in a short time scale a single particle is trapped among the neighboring particles in such a way that it maintains its original position almost unchanged and exchange kinetic energy mainly through phonon-like modes with the surrounding environment. On the other hand α relaxation corresponds to the slow rearrangements of the cages, that is the relaxation of the whole structure.

There is another way to understand this two step relaxation mechanism. One may assume the existence of many free energy minima separated by high energy barriers. During an aging process at early times the system is able to access at least part of the energy states by thermally overcoming barriers and getting out of local minima. However, as time passes, the system falls each time into a deeper minima which is more difficult to avoid. Consequently, the kinetics of the system during aging become slower and slower, in such a way that the system cannot reach the thermodynamic equilibrium. During aging, the viscosity of the system increases and the diffusion coefficient of the particles decreases. This suggests that the trapping of particles in cages formed by neighbors in real space is equivalent to the arrest of the system in a local minimum of the free energy landscape.

1.9 Definition of Various Non-ergodic States

1.9.1 Colloidal Gel

Colloidal gels can be thought of as a space filling or percolating network of particles [16, 23, 29, 31]. Particulate suspensions can form a percolating network when the system is destabilized. Under appropriate thermodynamic conditions, sufficient particle concentration and provided that the attraction between the particles are strong enough to induce aggregation, a space filling macroscopic structure is formed that effectively traps the solvent molecules [17, 22, 32]. Although not a thermodynamical equilibrium state, the percolating network undergoes a dramatic kinetic slowdown and the resulting gel has a long life [4, 13, 14, 15, 24]. The important parameter that affects this kinetic slowdown is the range of the attractive potential and its relation to the particle size and density [15, 27]. It is also important whether the contact of two particles results in a permanent bond, or if the particles have the additional freedom of rotating on top of each other. In the first case, the formation of the network is dominated by the diffusion [13] while in the second case there is a slow kinetic evolution driven by the phase separation occurring in the system [11, 28].

1.9.2 Colloidal Glass

Colloidal glasses are concentrated suspensions of microscopic particles in a liquid in which the particles' movements are constrained; they hold some freedom for local Brownian motions but are unable to diffuse over large lengths. Due to this localization, colloidal glasses at rest are amorphous solids. Nevertheless, they are typically soft solids, deforming elastically under small applied stresses, but yielding and flowing when stressed more strongly. A nice example is toothpaste which flows from the tube but sits as a solid piece of material on the brush [21]. Various natures of glasses like Wigner and attractive are present in the literature.

1.9.3 Comparison Between Gel and Glass

Both are disordered solid states, i.e. they do not flow if turned upside down, while their structural features display no long-range order. They can be differentiated, in the broader sense, by their density. Glasses typically arise in dense fluids, due to supercooling or compression, and the typical picture illustrating the mechanism of arrest is the so-called cage effect: particles are trapped by their nearest-neighbors and cannot relax to their underlying equilibrium state. Gels on the other hand are found at much lower densities and typically can be attributed to the formation of a particle network, where particles are linked together via attractive bonds. So while attraction is necessary to form a gel, a glass is typically repulsive and its colloidal prototype is the hard-sphere glass [23].

1.10 Structure of the Thesis

Having introduced the importance of investigating the complex phenomenology of colloids, I present here the outline of this thesis which deals with the experimental study of the dispersion stability, microstructure and phase transitions in anisotropic nanodiscs.

- Chapter 2 is dedicated to the materials and characterization techniques used to achieve the goal of the thesis. The anisotropic nano clays used were Laponite and Sodium Montmorillonite (Cloisite Na). The basic elements the characterization techniques were described in this chapter.
- Chapter 3 explores the phase diagram of aging Laponite in aqueous medium. Different phase states of Laponite system for the wide range of Laponite concentrations [0.1–3.5 % (w/v)] were established by suitable techniques.
- Chapter 4 describes the kinetics of orientational ordering of Laponite particles induced by water–air interface. Depolarized light scattering technique was employed to study the orientation and relaxation dynamics. A very fine tuned geometry was used to explore the same for the precise measurements.
- Chapter 5 investigates the phase diagram of aging Montmorillonite in aqueous medium. We have used various techniques to study the kinetics of gelation and to observe the equilibrium gels.
- Chapter 6 describes the sol behavior and gelation kinetics in mixed nanoclay dispersions of Laponite and Montmorillonite.
- Chapter 7 investigates the ergodicity breaking and aging dynamics in mixed nanoclay dispersions using light scattering experiment.
- Chapter 8 investigates the thermal induced irreversible ordering in mixed nanoclay dispersions. Landau theory of phase transitions was used to understand the thermal induced ordering.
- Chapter 9 deals with the aggregation and scaling behavior of nanoclays having different aspect ratio in alcohol solutions.
- Chapter 10 concludes and summarizes the main findings of this thesis, also describes a series of opportunities for pursuing interesting future work on similar systems.

References

1. Z. Cheng, J. Zhu, P.M. Chaikin, S.E. Phan, W.B. Russel, Nature of the divergence in low shear viscosity of colloidal hard-sphere dispersions. Phys. Rev. E **65**, 041405 (2002)
2. A.C.V. Coelho, P.S. Santos, Argilas especiais: o que são, caracterização e propriedades. Quim. Nova **30**, 146–152 (2007)
3. P.G. de Gennes, Soft matter. Sci. **256**, 495–497 (1992)

4. G. Foffi, C. De Michele, F. Sciortino, P. Tartaglia, Arrested phase separation in a short-ranged attractive colloidal system: a numerical study. J. Chem. Phys. **122**, 224903 (2005)
5. W. Gotze, L. Sjogren, Relaxation processes in supercooled liquids. Rep. Prog. Phys. **55**, 241 (1992)
6. T. Graham, Liquid diffusion applied to analysis. Phil. Trans. R. Soc. Lond. **151**, 183–224 (1861)
7. H.C. Hamaker, The London-van der Waals attraction between spherical particles. Physica **4**, 1058–1072 (1937)
8. M.D. Haw, Colloidal suspensions, Brownian motion, molecular reality: a short history. J. Phys.: Condens. Matter **14**, 7769–7779 (2002)
9. H.N.W. Lekkerkerker, Remco Tuinier, *Colloids and the Depletion Interaction* (Springer, New York, 2011)
10. R. Holyst, Some features of soft matter systems. Soft Matter **1**, 329–333 (2005)
11. S.W. Kamp, M.L. Kilfoil, Universal behaviour in the mechanical properties of weakly aggregated colloidal particles. Soft Matter **5**, 2438–2447 (2009)
12. Laponite—synthetic layered silicate—its chemistry, structure and relationship to natural clays. Laponite technical bulletin. 2003. L204/01 g, www.laponite.com
13. M.Y. Lin, H.M. Lindsay, D.A. Weitz, R.C. Ball, R. Klein, P. Meakin, Universality in colloid aggregation. Nature **339**, 360–362 (1989)
14. J.F.M. Lodge, D.M. Heyes, Structural evolution of phase-separating model colloidal liquids by Brownian dynamics computer simulation. J. Chem. Phys. **109**, 7567–7577 (1998)
15. J.F.M. Lodge, D.M. Heyes, Rheology of transient colloidal gels by Brownian dynamics computer simulation. J. Rheol. **43**, 219–244 (1999)
16. P.J. Lu, E. Zaccarelli, F. Ciulla, A.B. Schofield, F. Sciortino, D.A. Weitz, Gelation of particles with short-range attraction. Nature **453**, 499–504 (2008)
17. R. Mezzenga, P. Schurtenberger, A. Burbidge, M. Michel, Understanding foods as soft materials. Nat. Mat. **4**, 729–740 (2005)
18. M.C.D. Mourad, A.V. Petukhov, G.J. Vroege, H.N.W. Lekkerkerker, Lyotropic hexagonal columnar liquid crystals of large colloidal gibbsite platelets. Langmuir **26**, 14182–14187 (2010)
19. M.C.D. Mourad, D.V. Byelov, A.V. Petukhov, D.A.M. Winter, A.J. Verkleij, H.N.W. Lekkerkerker, Sol-gel transitions and liquid crystal phase transitions in concentrated aqueous suspensions of colloidal gibbsite platelets. J. Phys. Chem. B **113**, 11604–11613 (2009)
20. W.C.K. Poon, The physics of a model colloid-polymer mixture. J. Phys.: Condens. Matter **14**, 859 (2002)
21. P.N. Pusey, Colloidal glasses. J. Phys.: Condens. Matter **20**, 494202 (2008)
22. B. Rajaram, A. Mohraz, Microstructural response of dilute colloidal gels to nonlinear shear deformation. Soft Matter **6**, 2246–2259 (2010)
23. B. Ruzicka, E. Zaccarelli, L. Zulian, R. Angelini, M. Sztucki, A. Moussaïd, T. Narayanan, F. Sciortino, Observation of empty liquids and equilibrium gels in a colloidal clay. Nat. Mater. **10**, 56–60 (2011)
24. C.M. Sorensen, A. Chakrabarti, The sol to gel transition in irreversible particulate systems. Soft Matter **7**, 2284–2296 (2011)
25. T.F. Tadrosl, Correlation of viscoelastic properties of stable and flocculated suspensions with their interparticle interactions. Adv. Colloid Interface Sci. **68**, 97–200 (1996)
26. S.L. Tawari, D.L. Kochi, C. Cohen, Electrical double-layer effects on the brownian diffusivity and aggregation rate of laponite clay particles. J. Colloid Interf. Sci. **240**, 54–66 (2001)
27. V. Trappe, P. Sandkuhler, Colloidal gels—low-density disordered solid-like states. Curr. Opin. Colloid Interface Sci. **8**, 494–500 (2004)
28. V. Trappe, V. Prasad, L. Cipelletti, P.N. Segre, D.A. Weitz, Jamming phase diagram for attractive particles. Nature **411**, 772–775 (2001)
29. M.M. van Schooneveld, V.W.A. de Villeneuve, R.P.A. Dullens, D.G.A.L. Aarts, M.E. Leunissen, W.K. Kegel, Structure, stability, and formation pathways of colloidal gels in systems with short-range attraction and long-range repulsion. J. Phys. Chem. B **113**, 4560–4564 (2009)

30. E.J.W. Verwey, JTh Overbeek, *Theory of the Stability of Lyophobic Colloids* (Elsevier, Amsterdam, 1948)
31. E. Zaccarelli, Colloidal gels: equilibrium and non-equilibrium routes. J. Phys.: Condens. Matter **19**, 323101 (2007)
32. E. Zaccarelli, S.V. Buldyrev, E. La Nave, A.J. Moreno, I. Saika-Voivod, F. Sciortino, P. Tartaglia, Model for reversibile colloidal gelation. Phys. Rev. Lett. **94**, 218301 (2005)

Chapter 2
Materials and Characterization Techniques

Abstract This chapter reviews characterization techniques with emphasis on their introduction and principle of operation for the experiments to achieve the goal of this thesis. Lights scattering techniques like dynamic light scattering, static light scattering, depolarized light scattering; deformation technique like rheology, microscopic technique such as transmission electron microscopy were used to probe the samples. Other techniques such as Raman spectroscopy, viscometry and tensiometry were also used. In addition, we also give brief accounts of materials such as clays and their characterization and solvents used in this thesis work.

2.1 Materials Used

2.1.1 Nanoclays

Laponite RD and *Na* Montmorillonite (Cloisite *Na*) were purchased from Southern Clay products, USA and used as received. Sample preparation is described in respective chapters.

2.1.2 Solvents

Triple de-ionized water was purchased from Praveen Scientific, India. Organic solvents: methanol (MOH), ethanol (EOH), 1-propanol (POH), hexanol and octanol were purchased from Merck. The pH adjustments were done using 0.1N NaOH and 0.1N HCl obtained from Fisher Scientific, India. All solvents were of spectroscopic grade and used as received.

2.2 Characterization Techniques

2.2.1 Laser Light Scattering

Light scattering is a non-invasive technique for characterizing macromolecules and a wide range of particles in solution. In contrast to most methods of characterization, it does not require external calibration standards. In this sense, it is an absolute technique. Scattering is a very powerful technique to study the physical properties (structure and dynamics) of a system and is used routinely for measuring the hydrodynamic size of colloids and nanoparticles, particularly in dispersions. Scattering results from the interaction of the electrons in the molecules with oscillating electric field of radiation. Thus, a dipole is induced in the molecules, which oscillate with the electric field. Since an oscillating dipole is a source of electromagnetic radiation, the molecules emit light, called scattered light. The difference in energy between the incident and scattered radiation determines the nature of scattering whether the phenomena is elastic (Rayleigh scattering), quasi-elastic scattering (Rayleigh-Brillouin scattering) or inelastic scattering (Raman scattering). The technique of light scattering has an advantage of being non-invasive probe method provided the intensity is not high enough to ionize the system. To probe a wide range of length scale, different scattering mechanisms are used i.e. typical length scale accessible in light scattering is up to 1,000 nm whereas in the case of neutron scattering it is up to 100 nm, X-ray scattering can probe up to 100 nm and for probe length scale smaller than this, one has to go for electron scattering.

(i) **Dynamic Light Scattering**

Dynamic light scattering (DLS) probes the transport properties like diffusion coefficient which in turn determines the hydrodynamic radius of the nanoparticles, macromolecules etc. in liquid environments. In a light scattering experiment a monochromatic beam of laser light impinges on a sample and is scattered into a detector placed at an angle θ with respect to the transmitted beam. The intersection between the incident and scattered beams defines a volume V, called the scattering or illuminated volume [3]. Dynamic light scattering [4–8] is sensitive to the diffusion of scattering particles in solution, as it measures the intensity of light scattered at a fixed angle, which is then analyzed with an autocorrelator. The resulting correlation function has the particle diffusion coefficient as one of its arguments, which can be used to calculate the hydrodynamic radius of the particle through Stoke-Einstein relation.

In our work, dynamic laser light scattering experiments were performed on digital correlator (PhotoCor Instruments, USA) and a homemade goniometer that was operated in the multi-tau mode (logarithmically spaced channels). The time scale spanned 8-decades, i.e. from 0.5 μs to 10 s. The samples were loaded into 5 ml cylindrical borosilicate glass cells and sealed. These cells were housed inside a thermostated bath (scattering geometry), and the temperature was regulated by a PID temperature controller to an accuracy of ±0.1 °C. The excitation source was a He-Ne laser emitting at a wavelength of 632.8 nm in linearly polarized single frequency mode, which was

2.2 Characterization Techniques

focused on the sample cell and scattered light was detected by a photo-multiplier tube (Hamamatsu), and the signal was converted into intensity auto-correlation function by a digital correlator. The scattering angle was fixed at 90° and the measured intensity auto-correlation functions were analyzed by the CONTIN regression software after ensuring that the relaxation time distribution function did not contain more than one distribution. Robustness of the DLS results was decided based on two criteria: sample to sample accuracy, and data reproducibility within the same sample. In all the experiments, the difference between the measured and calculated base line was not allowed to go beyond ±0.1 %. The data that showed excessive baseline difference were rejected.

In a typical DLS experiment, we measured the time correlation function $g_2(t)$ of the scattered intensity $I(t)$ at a given q (scattering wave vector) defined by Berne et al. [4]

$$g_2(t) = \frac{\langle I(t') I(t'+t) \rangle}{\langle I(t') \rangle^2} \qquad (2.1)$$

where $q = (4\pi n/\lambda) \sin(\theta/2)$, n is the refractive index of the solution, θ is the scattering angle, and λ is the wavelength of light. The intensity correlation function $g_2(t)$ is related to the scattered field auto-correlation function, $g_1(t)$ through the Siegert relation [4–8]

$$g_2(t) = A + B|g_1(t)|^2 \qquad (2.2)$$

where A defines the baseline of the correlation function as,

$$|g_2(t)|_{t \to \infty} = A \qquad (2.3)$$

and B is the spatial coherence factor. The ratio (B/A) is the signal modulation and better data quality demands $(B/A) \geq 50$ %. For solutions containing particles undergoing Brownian motion (i.e., polymer or colloidal solutions), the field auto-correlation function, $g_1(t)$ is given as,

$$g_1(t) = \sum_i A_i \exp(-\Gamma_i t) \qquad (2.4)$$

where Γ_i is the relaxation frequency, which characterizes various relaxation modes that include relaxations due to the translational diffusion, rotational diffusion and bending modes etc. The relative mode strength (amplitude) of the ith relaxation mode is A_i. For the present case, center of mass diffusion is the dominant process and Γ_i has been identified as $\Gamma_i = D_i q^2$. Here the translational diffusion coefficient of the ith particle is D_i. The expression for $g_1(t)$ remains valid for polydisperse samples, and for situation where the relaxation frequency distribution has several peaks. Polydispersity P can be defined as,

$$P = \frac{\langle (D - \bar{D})^2 \rangle}{(\bar{D})^2} \qquad (2.5)$$

Further details about DLS can be found elsewhere [4, 8]. According to Einstein relation, the $\bar{D}_o$ (the z-average diffusion coefficient at infinite dilution) is inversely proportional to the translational frictional coefficient, f_t at infinite dilution given by the relation,

$$\bar{D}_o = \frac{k_B T}{f_t} \qquad (2.6)$$

where k_B is Boltzmann constant and T is absolute temperature. The value of f_t obtained via Eq. 2.6 can be used for a direct estimation of hydrodynamic radius, R_h of the particles provided they have a spherical shape using the relation $f_t = 6\pi\eta R_h$ as per the Stokes law. Further one can write

$$D = \frac{k_B T}{6\pi\eta R_h} \qquad (2.7)$$

where η is the solvent viscosity at temperature T. The intensity correlation function is related to diffusion coefficient of the particles and hence to the hydrodynamic radius by Stoke-Einstein equation through Eq. 2.7.

(ii) **Analysis of Non-ergodic Samples**

The aging behaviour and slow dynamics is best probed by dynamic light scattering experiments. The intensity auto-correlation function $g_2(q,t)$ cannot be correlated directly to the corresponding dynamic structure factor $g_1(q,t)$ via the Siegert relation $g_2(q,t) = A + B|g_1(q,t)|^2$, where A defines the baseline of the correlation function as $|g_2(t)|_{t\to\infty} = A$, and B is the spatial coherence factor. Complications are caused when the scattering centers in the dispersion phase are localized near fixed mean positions and are able only to execute limited Brownian motion about the same. Here the system is considered to be arrested. A completely ergodic system is one where the time and the ensemble averages are identical, and the system is stationary implying that the process is independent of the origin of time which is adequately satisfied only in case of scattering from homogeneous dilute solutions. The non-ergodicity problem in DLS measurements has been dealt with in several ways which include rotating the sample to probe the entire phase space, expanding the incident beam to increasing the scattering volume and extracting the non-ergodic contribution from the measured data as a heterodyne contribution.

Herein, we address the problem in the following way. The normalized intensity correlation function, $g_2(q, t)$, obtained from the sample in arrested phase can be related to the dynamic structure factor, $g_1(q, t)$ as [9]

$$g_2(q,t) = 1 + \beta'\left[2X(1-X)g_1(q,t) + X^2|g_1(q,t)|^2\right] \qquad (2.8)$$

where β' is the coherence factor having a maximum value of 1. In a real experiment it defines the signal modulation which is a measure of the signal-to-noise ratio in the data. The parameter X ($0 \leq X \leq 1$) defines the ergodicity via the amount of heterodyne contribution present in the correlation $g_2(q, t)$ data. When the value of $X = 1$, the system is completely ergodic i.e., in the sol state and the

Siegert relation is established; however in the arrested state $X < 1$ and the term $2X(1 - X)$ makes a finite contribution to $g_2(q, t)$ and hence it must be accounted for. The intercept of the plot of $[g_2(q, t) - 1]$ versus delay time t at $t \to 0$ gives $\beta' [2X - X^2]$ from which the value of X can be calculated if β' is known, which is an instrumental factor. The measured intensity auto-correlation data was analyzed exactly following the description given elsewhere [9]. The pre-factor of the linear term in $g_1(q, t)$ in Eq. 2.8 is much larger than the quadratic second term. Thus

$$g_1(q,t) \approx [g_2(q,t) - 1] / [2\beta'(X(1 - X))] \tag{2.9}$$

The values of the heterodyne parameter X is always less than 1 except in the homogeneous dilute solutions where ergodicity is expected. The exact evaluation of the dynamic structure factor $g_1(q, t)$ from the measured intensity auto-correlation function $g_2(q, t)$ was achieved by putting the values of β' and X into Eq. 2.9 and hence the non-ergodic contribution was removed.

(iii) Static Light Scattering

In static light scattering solute particles are taken as stationary. Intensity of the scattered light adjusted for background scattering and normalized to reference solvent gives the Rayleigh ratio $R_s(q)$ which can be expressed for dilute solution as [5–8]

$$\left(\frac{1}{R_s(q)}\right)\left(\frac{4\pi^2 n^2 (dn/dc)^2}{N_A \lambda^4}\right) = \frac{1}{P(q)M_w} + 2Bc \tag{2.10}$$

where c is concentration of macromolecule, N_A is Avagadro's number, M_w is the weight averaged molecular weight of macromolecule in solution, B is the second virial coefficient describing interparticle interactions in solution and $P(q)$ is a particle shape factor. If a particle size is small compared to light wavelength, then it obviously acts as point scatterer, and its shape is irrelevant for scattering i.e. $P(q) = 1$. When the particle size becomes comparable to the wavelength used in the experiment $P(q)$ can be approximated reasonably well as a quadratic function of qR_g and the scattering profiles can be used to determine the particle radius of gyration. For random coil in very low scattering angle limit (Guinier region), $P(q)$ is given by,

$$P(q) = 1 - \frac{q^2 R_g^2}{3}; \; Rg < q^{-1} \tag{2.11}$$

The static structure factor is the Fourier transform of the pair correlation function which in turn is the probability of finding of a particle at r, given that a particle is present at the origin. Figure 2.1 shows the schematic of the Light scattering set up used in my work for static and dynamic measurements. It comprises of a 35 mW He–Ne Laser radiating at 632.8 nm digital correlator (PhotoCor instrument, USA) that was operated in the multi-tau mode (logarithmically spaced channels). The time scale spanned 8-decades, i.e. from 0.5 μs to 10 s. The samples were loaded into cylindrical borosilicate glass cells and sealed. These cells were housed inside

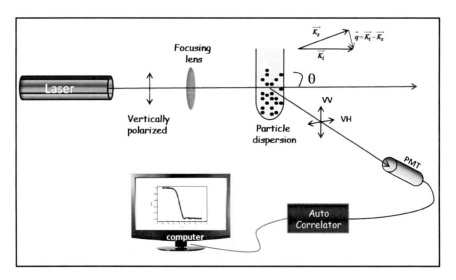

Fig. 2.1 Schematics of light scattering set up

a thermostated bath (scattering geometry) and temperature was regulated by a PID temperature controller to an accuracy of ±0.1 °C. The measured intensity autocorrelation function was analyzed by the CONTIN regression software.

(iv) **Depolarized Light Scattering**

The geometrical anisotropy of the scattering particles depolarizes the incident light, and the scattered electrical field can be decomposed into the parallel E_{VV} (q, t), and perpendicular E_{VH} (q, t) components, with respect to the direction of the incident polarization. These quantities fluctuate due to the random translational and rotational motion of the particles, and one can define two distinct dynamic structure factors or electric field correlation functions using these components of the scattered field. A block diagram for the depolarized light scattering is shown in Fig. 2.2. The analyzer was kept at right angle to the direction of the propagation of the laser light so as to minimize the effect of the stray light and the presence of an interference filter ensured that signal to noise ratio was robust. The angle in the analyzer was adjusted either to 0 or 90° respectively as per the requirement. The zero degree alignment of the analyzer meant that only the parallel component (I_{VV}) of the scattered light passed through it to the detector whereas the 90° alignment of the analyzer ensured passage of only perpendicular component (I_{VH}) of the scattered light.

2.2.2 Rheology

Colloidal systems can be viscoelastic, that is, they can show both solid and fluid like characteristics. The solid like response of the system is set by its elasticity

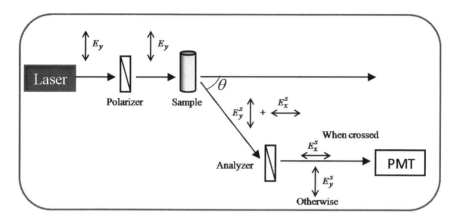

Fig. 2.2 Block diagram for depolarized light scattering

and the fluid like response by its viscosity. We can measure the material properties of the systems by means of rheometer. Rheology is the study of flow and deformation of matter. In practice, Rheology is principally concerned with extending the "classical" disciplines of elasticity and fluid mechanics (Newtonian) to materials whose mechanical behavior cannot be described with classical theories [7]. It is also concerned with establishing predictions for mechanical behavior (on the continuum mechanical scale) based on the micro or nanostructures of the materials. Rheology unites the seemingly unrelated fields of plasticity and non-Newtonian fluids by recognizing that both these types of materials are unable to support a shear stress in static equilibrium. In this sense, a plastic solid is a fluid. One of the tasks of rheology is to empirically establish the relationship between deformation and stress and their derivatives by adequate measurements. Such relationships are then amenable to mathematical treatment by the established methods of continuum mechanics.

There are two common approaches used in rotational Rheometer, controlled shear rate and controlled stress. In the controlled shear rate approach, the material being studied is placed between two plates. One of the plates is rotated at a fixed speed and the torsional force produced at the other plate is measured. Hence, speed (strain rate) is the independent variable and torque (stress) is the dependent variable. In the controlled stress approach, the situation is reversed. A torque (stress) is applied to one plate and the displacement or rotational speed (strain rate) of that same plate is measured. This latter approach (controlled stress) is better for determining apparent yield stress because the variable of primary interest can be more carefully controlled. That is, in the controlled stress approach it is possible to gradually increase the stress applied to the material and detect the point at which yield first occurs. Conversely, in the controlled rate approach, yielding actually has to be occurring before measurement can even occur. Hence, apparent yield stress can only be measured by back extrapolation from a finite level of motion to the point of zero motion [2].

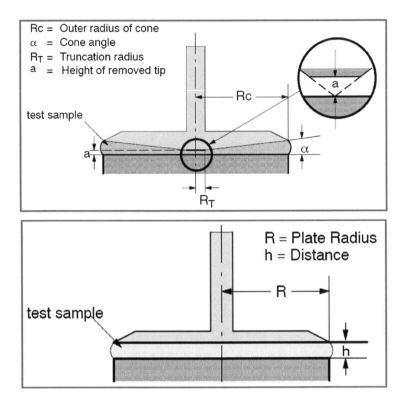

Fig. 2.3 Schematic of cone-plate and plate-plate geometry

The controlled stress approach provides three alternative experimental methods for determining apparent yield stress: (1) Stress can be ramped slowly from zero to some higher value. The stress level at which motion first occurs is the apparent yield stress. As indicated previously, this value may be affected by the stress ramp rate (rate of stress increase). (2) After initially shearing the material at a stress above the yield stress, the stress can be decreased in a slow ramp and the point where motion stops is the apparent yield point. Again as indicated previously, this value may be affected by the decreasing ramp rate and the time-dependent ability of the material to rebuild structure. (3) A creep experiment can be used where stress is applied to the material and strain (displacement) is monitored with time to establish an equilibrium yield stress.

The rheology experiments were performed using an AR-500 stress controlled rheometer (T.A. Instruments, UK). Different geometries used for rheological studies include the stainless steel cone plate geometry (diameter 20 mm and angle 20°) and parallel plate geometry (diameter 20 mm). The cone plate and parallel plate geometries are shown in Fig. 2.3. Rheological studies were carried out in different mode whose details are as follows:

2.2 Characterization Techniques

2.2.2.1 Modes of Operation

(a) **Flow Mode**

In flow mode, rheological measurements are normally performed in kinematic instruments in order to get quantitative results useful for design and development of products and process equipment. A rheometric measurement normally consists of a strain (deformation) or a stress analysis at a constant frequency combined with a frequency analysis, e.g. between 0.1 and 100 Hz. The fluid samples are normally divided into three different groups according to their flow behavior.

(i) **Newtonian Fluid:** It is a fluid whose stress vs. strain rate curve remains linear. The constant of proportionality is called viscosity (η). Shear viscosity remains constant with shear rate.

(ii) **Non-Newtonian Fluids, Time Independent $\eta = \eta(\gamma)$:** In Non-Newtonian Fluids, time independent case, the viscosity of fluid is dependent on shear rate but independent of the time of shearing. The viscosity is presented at a specific shear rate and referred to as the "apparent viscosity", "shear viscosity" or "shear dependent viscosity". Shear-thinning (a decrease in viscosity with increasing shear rate, also referred to as Pseudoplasticity) and Shear-thickening (an increase in viscosity with increasing shear rate, also referred to as Dilatancy) are the two properties of Non-Newtonian, time independent fluid.

(iii) **Non-Newtonian Fluids, Time Dependent $\eta = \eta(\gamma,t)$:** In Non-Newtonian Fluids, time dependent case, the viscosity of fluid is dependent on shear rate and the time during which shear rate is applied. Thixotropy (a decrease in apparent viscosity with time under constant shear rate or shear stress, followed by a gradual recovery, when the stress or shear rate is removed) and Rheopexy (an increase in apparent viscosity with time under constant shear rate or shear stress, followed by a gradual recovery when the stress or shear rate is removed, also called anti-thixotropy or negative thixotropy) are the two properties of Non- Newtonian, time dependent fluids.

(b) **Oscillation Mode**

(i) **Frequency Sweep Measurement:** The material response to increasing frequency (rate of deformation) is monitored at a constant amplitude (stress or strain) and temperature. The frequency sweep gives information of the elastic modulus G', the viscous modulus G'' and the phase angle 'δ'. A large value of G' in comparison of G'' indicates pronounced elastic (gel) properties of the sample being analyzed. For such a product the phase angle is also small, e.g. 20° (a phase angle of 0° and 90° means a perfectly elastic and viscous material respectively) shown in Fig. 2.4.

The frequency sweep gives information about the gel strength where a large slope of the curve G' vs omega indicates low strength and a small slope indicates high strength. The stress in a dynamic experiment is referred to as the complex stress σ^*. The complex stress can be separated into two components: an elastic stress in phase with the strain $\sigma' = \sigma^* \cos\delta$ (σ' is the degree to which material behaves like elastic solid) and

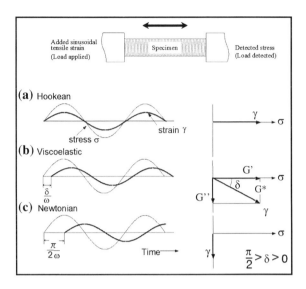

Fig. 2.4 Schematic picture of phase angle between stress (σ) and strain (γ) for **a** Hookean, **b** Viscoelastic and **c** Newtonian materials. Where G' is real or storage modulus G" is imaginary or loss modulus and G* is complex modulus

a viscous stress in phase with the strain rate $\sigma'' = \sigma^* \sin\delta$ (σ'' is the degree to which material behaves like a liquid). The complex modulus, $G^* = G' + iG'' = $ (stress*/strain) is the measure of overall resistance to deformation. The elastic (storage) modulus, $G' = $ (stress*/strain) $\cos\delta$ is the measure of elasticity of the material, is the ability of material to store energy. The viscous (loss) modulus $G'' = $ (stress*/strain) $\sin\delta$ is the ability of the material to dissipate energy as heat. Tan delta ($\tan\delta = G''/G'$) is the measure of material damping-such as vibration or sound damping.

The viscosity measured in an oscillatory experiment is a complex viscosity much the way the modulus can be expressed as the complex modulus. The complex viscosity contains an elastic component and a term similar to the steady state viscosity. The complex viscosity is defined as: $\eta^* = \eta' - i\eta''$ or $\eta^*(\omega) = G^*/\omega = \sqrt{|G'(\omega)^2 + G''(\omega)^2|}\big/\omega$. For viscoelastic gels, the complex viscosity is given by $|\eta*(\omega)| = \eta_0/\sqrt{1+\omega^2\tau_m^2}$ where, τ_m is the mean relaxation time for Maxwell model, η_0, the zero-shear viscosity related to the steady state viscosity and is the part of the measure of the rate of energy dissipation. Dynamic viscosity (η') for viscoelastic liquid approaches the steady flow viscosity (η_0) as the frequency approaches zero which is defined as $\eta' = G''/\omega$. The imaginary viscosity (η'') measures the elasticity or the stored energy and is related to the shear storage modulus which is defined as $\eta'' = G'/\omega$. The typical material response in a frequency sweep experiment is illustrated in Fig. 2.5.

(ii) **Temperature Sweep:** In our temperature sweep measurement, sample was loaded onto peltier plate of rheometer and allowed to equilibrate for 10 min. The periphery of geometry was coated with light silicon oil and enclosed by a wet sponge to minimize the solvent evaporation. Keeping the angular frequency fixed G' as a function of change in temperature of the peltier plate was recorded. The profile of G' versus temperature gives information about the thermal transition occurring in the material.

2.2 Characterization Techniques 27

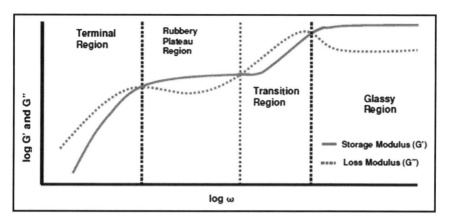

Fig. 2.5 The material response in a frequency sweep experiment

2.2.3 Viscometry

Viscosity measurements of nanoclay dispersions were performed on the Sine-wave Vibro (SV-10) Viscometer. SV-10 series has two thin plates as shown in Fig. 2.6. It drives with an ac frequency (30 Hz) by vibrating at constant sine-wave vibration in reverse phase like tuning fork. When the spring plates are vibrated with a same frequency, the amplitude varies in response to the amount of frictional force produced by the viscidity between the sensor plates and the sample.

To produce uniform amplitude, the vibro viscometer controls the electrical current that drives the vibration of the spring plates. Because the frictional force of viscidity is directly proportional to the viscosity, the driving electric current for vibrating the spring plates with a constant frequency to produce uniform amplitude is also directly proportional to the viscosity of each sample. The electromagnetic drive controls the vibration of the sensor plates to keep constant amplitude. The driving electric current which is the origin of exciting force, will be detected as the magnitude of the viscidity between the sensor plates and sample fluid. The vibro Viscometer measures the driving electric current, and then the viscosity is given by the positive correlation between the driving electric current and the viscosity. This resonating measuring and detection system has some advantage such as wide dynamic range and high resolution.

2.2.4 Tensiometry

Surface tension is a phenomenon in which the surface of a liquid, where the liquid is in contact with gas, acts like a thin elastic sheet. This term is typically used only when the liquid surface is in contact with gas (such as the air). If the surface is between two liquids such as water and oil, it is called interface. Energy is required

Fig. 2.6 Schematic of the vibro viscometer

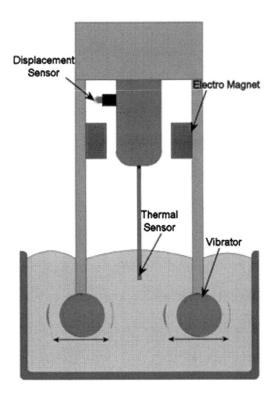

to change the form of this surface or interface. The work required to change the shape of a given surface is known as the interfacial or surface tension. Surface tension (denoted with the Greek variable γ) is defined as the ratio of the surface force F to the length L along which the force acts; $\gamma = F/L$. In order to consider the thermodynamics of the situation, it is sometimes useful to consider it in terms of work per unit area.

Tensiometer is an instrument to measure the surface tension and/or interfacial tension of a liquid using the plate, ring or drop methods. It is a more difficult experimental measurement to accomplish and is not nearly as accurate as the plate method. In plate method, we require a plate to contact the liquid surface. It is widely considered the simplest and most accurate method for surface tension measurement. Most of the KRUSS tensiometers determine the surface tension with the help of an optimally wettable probe suspended from a precision balance; this is either a plate or a ring. A height-adjustable sample carrier is used to bring the liquid to be measured into contact with the probe. A force acts on the balance as soon as the probe touches the surface. If the length of the probe is known (circumference of ring or length of plate) the force measured can be used to calculate the interfacial or surface tension. A further requirement is that the probe must have a very high surface energy. This is why a platinum-iridium alloy is used for the

2.2 Characterization Techniques

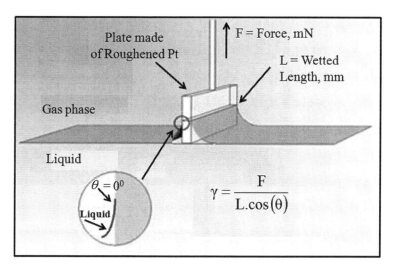

Fig. 2.7 Schematic of the plate method for measuring surface tension

ring and roughened platinum for the plate. In this work, K100MK2 KRUSS plate tensiometer, which is universal among tensiometer was used for surface tension measurement.

Figure 2.7 illustrates the schematic of plate method. In the plate method, the liquid is raised until the contact between the surface or interface, and the plate is established. Maximum tension acts on the balance at this instant; this means that the sample does not need to be moved again during the measurement. The tension is calculated using the following equation.

$$\gamma = \frac{F}{L \cdot \cos(\theta)} \qquad (2.12)$$

where γ is surface tension; F is force acting on the balance; L is wetted length and θ is contact angle. The plate is made of roughened platinum and is optimally wetted so that the contact angle θ is virtually 0°. This means that the term $\cos \theta$ has a value of approximately 1, so that only the measured force and the length of the plate need to be taken into consideration.

2.2.5 Electrophoresis

Electrophoresis is the phenomenon of the movement of a charged particle relative to the liquid suspended in, under the influence of an applied electric field [12–15]. The particle surface charge is one of the factors determining the physical stability of colloidal suspension and emulsions. Surface forces at the interface of

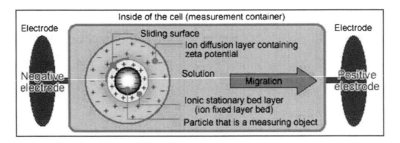

Fig. 2.8 The picture on the top shows a negatively charged molecule. Due to electrostatic attraction the counter ions make a layer around it to neutralize the interfacial potential, called stern layer or fixed layer and the ions with a small amount of the same polarity are distributed diffusely, called diffuse layer. (from Microtec Co., Ltd.)

the particle, and the liquid are very important because of the microscopic size of the colloids. Each colloid carries a like electrical charge which produces a force of mutual electrostatic repulsion between adjacent particles. Therefore, surface charge is a very good index of the magnitude of the interaction between colloidal particles, and this measurement is an important parameter to characterize colloidal dispersions. Typically, the particle charge is quantified as the so called zeta potential, which is measured e.g. via the electrophoretic mobility of the particles in an electric field. Alternatively the particle charge can be quantified in surface charge per surface unit, determined by colloid titration. It can be defined as the measure of overall charge, a particle acquires in a specific medium. In this thesis, the charge is characterized by the zeta potential.

The zeta potential theory is described in very detail by Müller in [11]. The development of a net charge at the particle surface affects the distribution of ions in the surrounding interfacial region, resulting in an increased concentration of counter ions (ions of opposite charge to that of the particle) close to the surface. Thus an electrical double layer exists around each particle as shown in Fig. 2.8. The liquid layer surrounding the particle exists as two parts; an inner region, called the Stern layer, where the ions are strongly bound and an outer, diffuse, region where they are less firmly attached. Within the diffuse layer there is a notional boundary inside which the ions and particles form a stable entity. When a particle moves (e.g. due to gravity), ions within the boundary move with it, but any ions beyond the boundary do not travel with the particle. This boundary is called the surface of hydrodynamic shear or slipping plane. The potential that exists at this boundary is known as the Zeta potential. That is to say, that the potential at the surface of shear of a charged particle i.e. for macro-ions it is the potential at the surface of the hydrodynamic particles. The conventional symbol used for zeta potential is "ζ" and it is usually measured in milli-volts (mV).

A zeta potential value quoted without a definition of its environment (pH, ionic strength, concentration of any additives) is a meaningless number. The

2.2 Characterization Techniques

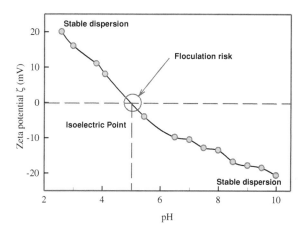

Fig. 2.9 Zeta potential of Gelatin B as function of pH. The system responds to pH by becoming more positively charged at lower pH (positive zeta potential) and more negatively charged at higher pH (negative zeta potential). At some intermediate pH, zeta potential is zero called the iso-electric point

magnitude of zeta potential gives an indication of the potential stability of the colloidal system. As the zeta rises, the repulsion between the particles becomes stronger. The stability of the dispersion gets higher while, on the other hand, as zeta approaches zero, the electrostatic build-up is relaxed allowing easier aggregation. A dividing line between the stable and unstable aqueous dispersion is generally taken at either +30 or −30 mV. Particles with zeta potentials more negative than −30 mV or more positive than +30 mV are normally considered stable (see Fig. 2.9). The most important factor that affects zeta potential is pH. The pH at which the net surface charge on the molecules became zero (or at which zeta potential becomes zero) is known as the isoelectric point of the solution. It is normally the point where the colloidal system is least stable. In general, a zeta potential versus pH curve will be positive at low pH and lower or negative at high pH.

We performed electrophoresis measurement on a zeta potential instrument (ZC-2000, Microtec, Japan). The sample solution is always necessary to dilute with dispersion medium (solvent). Dilution was done to isolate all individual particles from the aggregates and to know the distribution of charges on the surfaces of the single particle. In order to minimize the influence of electrolysis to the measurements, molybdenum (+) and platinum (−) plates were used as electrodes. Also, during the measurements, the cell chamber tap on molybdenum electrode was kept open to release the air bubbles, for reducing their effects on the particle movement. During the measurements, the molybdenum anode was cleaned each time as it turned from a metallic to blue-black color. The instrument was calibrated against 10^{-4} m AgI solution to meet the pre-measurement conditions set by the manufacturer. The nominal distribution of zeta potential is expected to be in the range −40 to −50 mV. To a good approximation, it can be presumed that zeta potential is directly proportional to the net charge on the particle if one assumes that the particles are non free-draining. Thus, the net surface charge density of the clusters becomes amenable.

2.2.6 Raman Spectroscopy

The energy of most molecular vibrations corresponds to the infrared region of the electromagnetic spectrum, and these vibrations may be detected in the range of few hundred to a few thousands cm^{-1} and measured in a Raman or infrared (IR) spectrum. Thus, Raman spectroscopy provides information about molecular vibrations that can be used for sample identification and quantitation. In Raman scattering measurements, a monochromatic light source, usually a single mode laser is incident on the sample and scattered light is detected. The majority of the scattered light is of the same frequency as the excitation source; this is known as Rayleigh or elastic scattering. Very small amount of the scattered light with lower (Stokes lines) or higher (Anti-Stokes lines) wavelength is shifted in frequencies from the laser frequency due to interactions between the incident electromagnetic waves, and the vibrational energy levels of the molecules in the sample. The energy difference between the incident and scattered light, so called Raman shift and usually expressed in terms of wave number cm^{-1} = $1/\lambda$. This shift is independent of the frequency of the incident light. It is just a function of the properties of the scattering molecule. Using classical considerations the Stokes and Anti-Stokes lines should possess the same intensity but quantum-theoretical considerations show that the Raman effect is based on an inelastic photon scattering, which starts at the Stokes lines on a level with a small vibration quantum number and ends on a vibration level with a higher quantum number, whereas with the Anti-Stokes effect, the scattering process shows the inverse behavior.

When monochromatic light of frequency ν_P strikes a cell containing a transparent substance, most of the light passes through unaffected. However, some of the light is scattered by the sample in all directions. The scattered light contains photons having different frequencies from that of the incident light such as ($\nu_P - \nu_{vib}$) and ($\nu_P + \nu_{vib}$) and the same frequency ν_P as the incident light (elastic scattering). The difference corresponds to the energy change which has taken place within the molecule. The lines having lower frequency than the incident light ($\nu_P - \nu_{vib}$) are termed as Stokes lines (red shifted), while the high-frequency lines ($\nu_P + \nu_{vib}$) are termed as anti-Stokes lines (blue shifted). Figure 2.10 illustrates the energy level diagram in Raman scattering. For vibrational transitions, the anti-Stokes lines are usually weaker than the Stokes lines. In contrast to infrared spectroscopy, Raman spectroscopy is dependent on a change in the polarizability of a molecule during the vibration, and the intensity is related to the polarizability of the vibrating atoms, and their bonds present [14].

The water structure in clay dispersions were best probed by Raman spectroscopy as the Raman active vibrational modes are very sensitive to the local environment. The Raman spectra were recorded in back scattering configuration using a Renishaw Raman microscope with Ar-ion laser excitation at 514 nm. An incident maximum laser power of 50 mW was applied in order to avoid peak shifts due to thermal heating or structure transformations during data acquisition. We have also recorded Raman spectrum on a FT-IR/Raman Spectrometer with Microscope-Varian 7,000 FT-Raman with Ar-ion laser excitation at 1,064 nm and 450 mW power.

2.2 Characterization Techniques

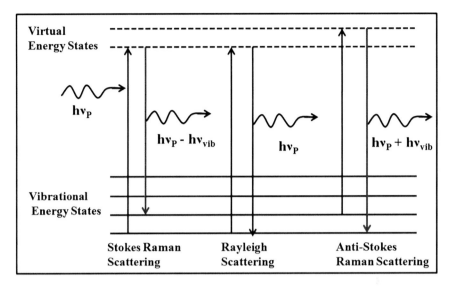

Fig. 2.10 Energy level diagram of states in Raman signal

2.2.7 Transmission Electron Microscopy

Transmission Electron Microscopy (TEM) is an imaging technique, uses a stream of monochromatic electrons that are focused to a small, thin, coherent beam by the use of condenser lenses. The wavelength of an electron is described by the de Broglie expression $\lambda = h/mv$, thus with suitable acceleration the desired angstrom wavelengths are achieved. In a transmission electron microscope (TEM), the ray of electrons is produced by a pin-shaped cathode heated up by current. The electrons are vacuumed up by a high voltage at the anode. The acceleration voltage is between 50 and 150 kV. The higher it is, the shorter are the electron waves, and the higher is the power of resolution. However, this factor is hardly ever limiting. The power of resolution of electron microscopy is usually restrained by the quality of the lens-systems and especially by the technique with which the preparation has been achieved. Modern gadgets have powers of resolution that range from 0.2 to 0.3 nm. The useful magnification is therefore, around 2,80,000×. The accelerated ray of electrons passes a drill-hole at the bottom of the anode. Its working principle is analogous to that of a ray of light in a light microscope. The lens-systems consist of electronic coils generating an electromagnetic field. The ray is first focused by a condenser. It then passes through the object, where it is partially deflected. The degree of deflection depends on the electron density of the object. The greater the mass of the atoms, the greater is the degree of deflection. After passing the object the scattered electrons are collected by an objective. Thereby, an image is formed, that is subsequently enlarged by an additional lens-system (called projective with electron microscopes). Thus formed image is made

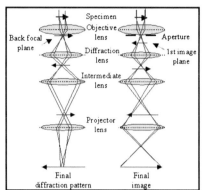 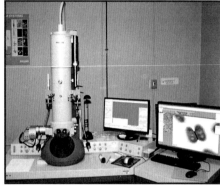

Fig. 2.11 Fei-Philips Morgagni 26&D transmission electron microscope used in the present work

visible on a fluorescent screen, or it is documented on photographic material. Photos taken with electron microscopes are in always black and white.

The TEM is an effective method to determine and/or verify the size and shape uniformity of nanoparticles within a sample, and used heavily in both material science/metallurgy and the biological sciences. In both cases, the specimens must be very thin and able to withstand the high vacuum present inside the instrument. In this work, the average particle size and morphology of clays were examined by Fei-Philips Morgagni 26&D transmission electron microscope (Digital TEM with image analysis system and Maximum Magnification = 2,80,000×) operating at a voltage of 100 kV (see Fig. 2.11). The aqueous dispersion of the clays was drop-cast onto a carbon coated copper grid and grid was air dried at room temperature (20 °C) before loading onto the microscope. In analytical TEMs the elemental composition of the specimen can be determined by analysing its X-ray spectrum or the energy-loss spectrum of the transmitted electrons. Modern research TEMs may include aberration correctors, to reduce the amount of distortions in the image, allowing information on features on the scale of 0.1 nm to be obtained (resolutions down to 0.08 nm has been demonstrated, so far). Monochromators may also be used, which reduce the energy spread of the incident electron beam to less than 0.15 eV.

2.2.8 X-Ray Diffraction

X-ray powder diffraction (XRD) is a convenient and powerful technique for materials investigation and was used for evaluating crystal structure and purities examination [1, 10]. For X-ray Diffraction (XRD) applications, the wavelength of X-rays roughly in the range between 0.01 and 10 nm (1–120 keV) are used. Because the wavelength of X-rays is comparable to the size of atoms, they are ideally suited for probing the structural arrangement of atoms and molecules in a wide range of materials [1, 10]. In X-rays diffraction experiment, an X-ray tube generates X-rays

Fig. 2.12 Schematic of X-ray diffraction

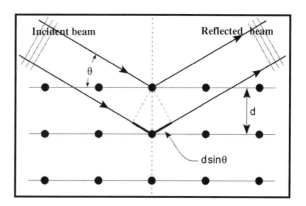

by focusing an electron beam that has been accelerated across a high voltage field and bombards a stationary or rotating solid target. As electrons collide with atoms in the target and slow down, a continuous spectrum of X-rays are emitted, which are termed Bremsstrahlung radiation. The high energy electrons also eject inner shell electrons in atoms through the ionisation process. When a free electron fills the shell, an X-ray photon with energy characteristic of the target material is emitted. Two common targets are Mo and Cu, which have strong K_α X-ray emissions at 0.71073 and 1.5418 Å, respectively. W. L. Bragg found an simple geometrical interpretation $n\lambda = 2d \sin\theta$ (λ is the wavelength of incident waves which have an angle of incidence θ to a set of lattice planes at distance d, n is an integer), this equation making it easy to relate the angle of diffraction to the interplanar spacing and to allow us to make accurate quantification of the results of experiments carried out to determine crystal structure (see Fig. 2.12). In typical measurements, the diffractometer projects a beam of X-rays through the crystal. The diffracted beam is collimated through a narrow slit and passed through a nickel filter; in order to ensure that only one wavelength of X-rays reaches the counter.

In our work, the X-ray diffraction (XRD) measurements were recorded with a PANalytical X'Pert PRO diffractometer using a solid state detector with a monochromatized Cu $k_{\alpha 1}$ (λ_{Cu} = 1.54060 Å) radiation source at 45 kV. The powder sample was put on a powder holder. When the Bragg conditions for constructive interference are obtained, a "reflection" is produced, and the relative peak height is generally proportional to the number of grains in a preferred orientation. The x-ray spectra generated by this technique provide a structural finger-print of the unknown crystalline materials. Mostly, the diffraction data is used for phase identification. Each crystalline powder gives a unique diffraction diagram, which is the basis for a qualitative analysis by X-ray diffraction. Identification is practically always accompanied by the systematic comparison of the obtained spectrum with a standard one (a pattern), taken from any X-ray powder data file catalogues, published by the American Society for Testing and Materials (JCPDS). The diffraction profiles of a mixture of crystalline specimens consist in the spectra of each of the individual crystalline substances present, superposed. Thus, by examining the diffraction pattern, one can identify the crystalline phase of the material.

References

1. L.E. Alexander, *X-Ray Diffraction in Polymer Science* (Wiley, London, 1969)
2. H.A. Barnes, *A Handbook of Elementary Rheology* (University of Wales, Institute of Non-Newtonian fluid Mechanics, Aberystwyth, 2000)
3. G.B. Bendek, in *Thermal fluctuations and scattering of light; Lectures at Brandeis Summer Institute for Theoretical Physics* (1968)
4. B.J. Berne, R. Pecora, *Dynamic Light Scattering with Applications to Chemistry, Biology, and Physics* (Wiley Interscience, New York, 1976)
5. W. Burchard, in *Laser Light Scattering in Biochemistry*, ed. by S.E. Harding, D.B. Sattelle and V.A. Bloomfield (Royal Society of Chemistry, Cambridge, 1992)
6. W. Burchard, M. Schmidt, W.H. Stockmayer, Information on polydispersity and branching from combined quasi-elastic and intergrated scattering. Macromolecules **13**, 1265 (1980)
7. N.P. Cheremisinoff, P.N. Cheremisinoff, Introduction to Polymer Rheology and Processing. ISBN 13:9780849344022, CRC Press (1993)
8. B. Chu, *Laser Light Scattering* (Academic Press, New York, 1974)
9. T. Coviello, E. Geissler, D. Meier, Static and dynamic light scattering by a thermoreversible Gel from Rhizobium leguminosarum 8002 Exopolysaccharide. Macromolecules **30**, 2008 (1997)
10. H.P. Klug, L.E. Alexander, *X-Ray Diffraction Procedures for Polycrystalline and Amorphous Materials* (Wiley, New York, 1954)
11. R.H. Müller, Zetapotential und Partikelladung in der Laborpraxis—Einführung in die Theorie, praktische Meßdurchführung, Dateninterpretation, Wissenschaftliche Verlagsgesellschaft Stuttgart, 254 S (1996)
12. H. Ohshima, Electrostatic interaction between two spherical colloidal particles Adv. Coll. In. Sci. **53**, 77–102 (1994)
13. H. Ohshima, Electrophoresis of soft particles. Ad. Coll. In. Sci. **62**, 189–235 (1995)
14. B.P. Straughan, *Spectroscopy*, vol. 2. (Chapman & Hall, 1976). ISBN 0412133709, 9780412133701
15. M. von Smoluchowski, Versuch einer mathematischen theorie der Koagulationskinetik kolloidaler Lösungen. Z. Phys. Chem. **92**, 129–168 (1918)

Chapter 3
Phase Diagram of Aging Laponite Dispersions

Abstract This chapter presents the age dependent hydration, heterogeneity and network rigidity of aqueous Laponite dispersions in its sol, gel and glass phase. Time dependent 3-D phase diagram of Laponite dispersions was proposed. The 3-D phase diagram is based on the hydration of the extensively aged Laponite dispersions, scattering techniques and the rheological studies.

3.1 Introduction

Laponite suspensions exhibit non trivial aging dynamics which is evident from numerous studies undertaken in the past Ruzicka et al. [31] have clearly shown that formation of gel and glass-like self-assembled structures are intrinsic to the aqueous dispersions of this clay. In general, dynamical arrest in soft colloidal systems has recently become the subject of intense research activity. The fine tuning of the control parameters allows the possibility to tailor the macroscopic properties of the resulting non-ergodic states. Several mechanisms of dynamical arrest have been identified and a rich phenomenology has been predicted both at high (re-entrant liquid-glass line, attractive and repulsive glasses) and low clay concentrations where gelation was shown to occur from different thermodynamic routes [38]. Recently, the existence of equilibrium gels in systems with intrinsic anisotropy (patchy colloids) has been predicted from experimental and numerical studies [30, 38]. The relevance of anisotropic interactions in colloidal systems has also recently emerged in the context of the rational design of novel soft materials. These patchy colloids are in fact today considered to be the novel building blocks of a bottom–up approach towards the realization of self-assembled bulk materials with pre-defined properties [13]. The aging of the Laponite dispersions has been most interesting subject in the literature [1, 6, 7, 17–19, 25, 32, 34].

The interaction between water and charged particles is an issue that is at the center of understanding of many biological processes [3]. Biological systems comprise charged particles and complex macromolecules such as proteins, nucleic acids, polysaccharides etc. in aqueous solutions as essential constituents. A hypothesis has been proposed that during the origin of life the primitive cells assembled in hydrogel environment [35]. According to this hypothesis the ensemble of charged particles, macromolecules and water was assumed to form a gel which essentially defined the early cell. Thus, a gel-like precursor assembly of a cell can function on a primitive basis without the necessity of the presence of a cell membrane. More generally, it is assumed that the biological function of proteins and other intracellular macromolecules cannot be understood without taking into account the mobility of the water molecules surrounding them in vivo [3]. Further, ion state of the aqueous solution has an influence on the biological function of proteins [15]. The mobility of water on inter and intra molecular length scales and the nature of water molecules have been reported in different systems [12, 16, 33] used time-of-flight neutron scattering to study the rotational and translational diffusion of water in Ca- and Na-montmorillonite. Other studies like diffusion of water and the dependence of temperature, and hydration in clay systems through experiments and simulations have been reported. Desai et al. [11] found that the included water was likely to cause the phases in the bentonite clays. Thus the study reported in this chapter is related to the hydration of Laponite and its phase behavior which was studied with aging.

A pertinent question arises; does the amount of structured water vary with the concentration of clay particles and aging, and how it impacts the dispersion phase? This required a systematic and controlled mapping of macroscopic structural properties of clay dispersions during their aging process. This constituted the principal objective of the work presented in this chapter and in addition, a 3-D time-dependent phase diagram for Laponite system has been proposed herein.

3.2 Sample Preparation

Laponite RD clay was purchased from Southern Clay Products, USA and used as received. We dried the Laponite clay for 4 h to remove the moisture present. Laponite dispersions were prepared as described in [30]. Stock dispersion was prepared by dissolving the powder clay in deionized water at room temperature. The dispersions were filtered through 0.45 μm Millipore filter and the pH was adjusted to 10. This time point was defined as $t_w = 0$, where t_w is called waiting time. All experiments were performed under room temperature conditions, and relative humidity <50 %.

3.3 Results and Discussion

3.3.1 Hydration of Laponite

All the clays are capable of adsorbing water onto their surface. The presence of adsorbed water covering the clay particles produces characteristic cohesive plastic behavior of clay minerals. The water adsorbed onto the clay surface has properties

3.3 Results and Discussion

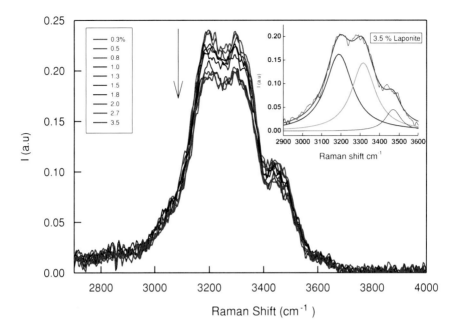

Fig. 3.1 Raman spectra of Laponite dispersions recorded at 25 °C. Note the considerable change in peak heights in the 2,900–3,600 cm^{-1} region of the spectrum where the signature O–H stretching vibrations are located. *Arrow* indicates direction of the spectrum with increasing the concentration of clay. These peaks were deconvoluted to resolve the individual peaks. *Inset* shows deconvolution of raw Raman spectrum of 3.5 % (w/v) dispersions. The best deconvolution result was observed for three component Lorentzian fitting where $\chi^2 > 0.99$

between bulk liquid water and ice, and forms a primary structured water layer and a more diffusion-less water layer. The thickness of water layer varies depending on the clay surface charge, the exchange cation and cation hydration and the salinity of the aqueous solution. It is also mentioned in the literature that the presence of charge on the laponite surface drives the water adsorption toward perpendicular and tilted configurations, in stark contrast to the uncharged surface, where the only adsorption mode consist of water molecules lying parallel to the surface [2]. The nature and behavior of water adsorbed onto the inter layers of clay mineral is a function of (i) polar nature of the water molecules (ii) the size of the inter layer cations and their hydration state and (iii) the location and value of the charge on the silicate layers of clay mineral structures [14, 19–23]. Sodium dominated clay shows a substantial swelling in pure water and it undergoes sol-gel transition [19–23].

The water structure in clay dispersions is best probed by Raman spectroscopy as the Raman active vibrational modes are very sensitive to the local environment. The raw Raman spectra obtained from clay dispersions are shown in Fig. 3.1 for samples having various solid concentrations. The data shown in Fig. 3.1 is restricted to the frequency domain 2,900–3,600 cm^{-1} because the OH-stretch modes of water reside in this regime.

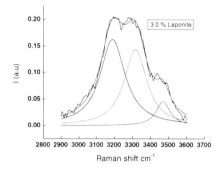 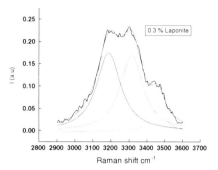

Fig. 3.2 Representative deconvolution of raw Raman spectra of 0.3 and 3.0 % (w/v) Laponite dispersions. The best result was observed for three component Lorentzian fitting where $X^2 > 0.99$ for all the cases

Raman spectra were fitted to 3-component Lorentzian functions using Origin 6.1 software, which was robust and highly reproducible with the overall statistical error remaining less than typically 2–3 % (Fig. 3.1 inset). Similar deconvolution for 0.3 and 3.0 % (w/v) Laponite dispersions are shown in Fig. 3.2. For the peaks located at 3,200, 3,300 and 3,400 cm^{-1} the chi-squared value obtained was 0.995 for water. Thus, the spectral analysis was robust which made us believe that the peaks in question were resolvable though they were separated by small Raman shift. This procedure yielded three discernible bands in the frequency range of 3,000–3,600 cm^{-1} and these deconvoluted peaks (3,200, 3,310 and 3,460 cm^{-1}) are identified as the characteristic vibrational modes of water [8, 36]. The in-phase vibration of OH stretching mode generates the peak at 3,200 cm^{-1}. This signifies the structurally arranged water, also referred to as ice like structures. The fraction at 3,310 cm^{-1} originates from the partial- structured water. The peak around 3,460 cm^{-1} arises from amorphous water or bulk water (poorly H-bonded water molecules). The three Raman components whose central frequencies correspond to about 3,200, 3,310, and 3,460 cm^{-1} collectively summed together, as described above, were used to define the hydrogen bonding configuration in the dispersion.

Figure 3.3a and b illustrate the variation in the amount of structured (3,200 cm^{-1}) and partially structured (3,310 cm^{-1}) water with the concentration and aging of dispersions which is quite revealing. A closer perusal of the data shown in Fig. 3.3a reveals three identifiable region: (i) for $c < 0.8$ % (w/v) the amount of structured water decreased by as much as 20 % for samples with higher solid content and for a given clay concentration this decrease was close to 12 % (ii) in the second region ($0.8 < c < 2.0$ % (w/v)), the structured water concentration remained independent of clay concentration and (iii) in the third region ($c > 2$ % (w/v)), the content of structured water increased with clay concentration. It is to be noted that the data pertain to three distinct set of measurements that was taken on the nascent ($t_w = 1$ h), less aged (5 months) and fully aged samples (9 months and 1 year). It is interesting to note that all these measurements consistently adhered to the qualitative category wise description already defined.

3.3 Results and Discussion

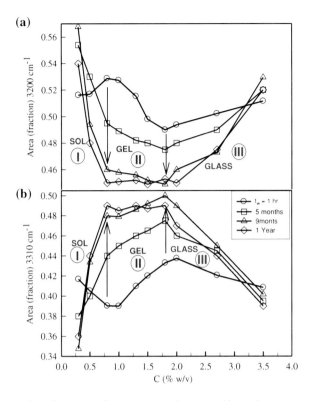

Fig. 3.3 The variation of the area of the **a** structured water and **b** semi structured water with the concentration of the clay at different aging times as mentioned in the graph. *Arrows* indicate the aging direction at sol-gel and gel-glass transitions. The size of the symbols indicates the error in the measurement

Thus, based on this data and reports available in the literature, it can be concluded that region-I was a sol state, II was a gel phase and III was a glassy formation. These three phases had their signature hydration characteristics with the gel seem to be deficient in structured water content. Interestingly, our measurements are in agreement with the results reported by [29], where they found the phase separation terminates at 1 % (w/v). Although, the sol state was monitored for up to 1 year, and the phase separation occurs at much longer timescales, so it may not be a complete or true sol state, but a phase that will eventually form dilute sol and a more dense gel. Dilution test was carried out on the samples to ensure the distinction between phases II and III [31]. Interestingly, the amount of structured water found in glassy phase was not too different from that of the sol state.

Figure 3.3b data complements this conclusion as it gives information about the semi or partially structured water, which also exhibits similar behavior as was seen in the case of structured water. The loss in the structured water was gained as the semi structured water. One report was found in the literature where similar observation was made [24].

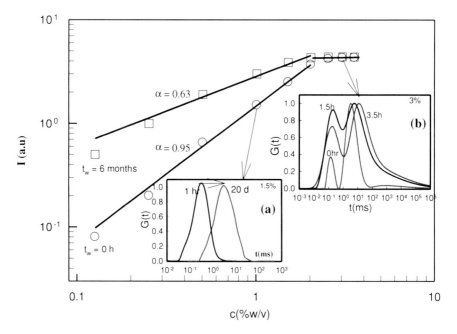

Fig. 3.4 The variation of intensity of light scattered from the dispersions of clay samples measured at room temperature. The power-law exponents (α) are indicated in the figure for each curve. The *solid lines* are least-squares fitting of data to power-law: I (q, c) ~ c^α. Inset **a** shows the distribution of relaxation time for c = 1.5 % (w/v) at t_w = 1 h and 20 days, and inset **b** shows distribution of relaxation time for c = 3.0 % (w/v) at t_w = 0, 1.5 and 3.5 h

3.3.2 Growth of Structures

The time-dependent internal self-assembly of colloidal disks of Laponite in the continuous phase attributes unique and distinct concentration dependent phase characteristics to these dispersions. This was examined by light scattering experiments that probed the samples over a typical length scale ~λ/n where λ is wavelength of excitation source and n is refractive index of dispersion medium [5]. We observed that the variation in the intensity of the light scattered off these dispersions with concentration up to c = 2 % (w/v) followed a power-law,

$$I(q, c) \sim c^\alpha \qquad (3.1)$$

where α = 0.95, and it remained constant for c > 2 % (w/v).

The static light scattering data is shown in Fig. 3.4. It is interesting to note that light scattering data does not distinguish between a sol and a gel state. In the concentration regime below 2 % (w/v), the power-law exponent changed from α = 0.95 (nascent sample) to 0.63 (6 months sample) with the aging. The intensity profile was invariant of concentration beyond 2 % (w/v) which was a signature of saturation in structure formation, manifestation of a hydrated amorphous phase (glassy state). Note that in this concentration regime the structured water presence was comparable to that of the sol state (Fig. 3.4a). Similar intensity profile was observed by Bellour et al. [4], but the aging effect was either not reported or observed.

Dynamic light scattering probing was used to observe the distribution of relaxation times in different phases of the Laponite suspension. A single relaxation (fast mode) was observed in case of gels in the range $1 < c < 2$ % (w/v), as shown in Fig. 3.4a, which means the network itself undergoes Brownian motion. In case of suspensions above $c = 2$ % (w/v) bimodal distribution of relaxations was characterized by a fast and a slow mode as shown in Fig. 3.4b. The fast mode corresponds to the collective Brownian motion of self-assembled particles as a single entity in this case. The second relaxation time corresponds to the time taken by the individual particles and their small aggregates to move out of cage formed of large self-assembled structures. Here again the fast mode component did not reveal any age dependence whereas the slow mode relaxation shifted to longer time scales implying dynamic arrest of the system.

3.3.3 Viscoelastic Behaviour

Colloidal gels and glasses are well known for their unique rheological attributes. We started from a routine amplitude sweep experiment where a sinusoidal deformation was imposed on the sample, and its amplitude was ramped logarithmically at a constant frequency. For structured materials, the elastic storage modulus G' and the viscous modulus G'' typically remain constant up to a limiting deformation or stress value. The stress at which the viscous modulus overtakes the elastic modulus is considered as the yield stress of the sample.

Yield stress σ and storage modulii G' measurements show that there are two power-law relations governing their dependence on clay concentration which is illustrated in Fig. 3.5.

$$G' \sim c^{\delta_1}; \quad \sigma \sim c^{\delta_2} \tag{3.2}$$

The first dependence occurs above 2 % (w/v) with the exponent $\delta_1 = 3 \pm 0.3$ which corresponds to G' and $\delta_2 = 2.8 \pm 0.3$, which corresponds to σ, and result was consistent with the observation made by Pignon et al. [26]. The second dependence occurs for the gel samples for $1 < c < 2$ % (w/v) with the power-law exponent $\delta_1 = 2.3 \pm 0.3$, for G' and $\delta_2 = 2.4 \pm 0.3$ for σ, which confirmed the existence of two clearly identified regions in the dispersions in the concentration regime above 1 % (w/v). In order to obtain further insight into the viscoelastic properties of the samples, the loss tangent characteristics at various clay concentrations was examined and the same data is presented in Fig. 3.6.

The tangent of the phase angle is the ratio of the loss modulus and the storage modulus, $\tan \delta = G''/G'$ which is a measure of the damping ability of the material. It was found that the loss tangent decreased with the frequency for all samples.

Fig. 3.5 Evolution of yield stress and the storage modulus as a function of clay concentration. The exponents are indicated for each curve. *Open circles* indicate the elastic modulus and open squares indicate the yield stress. Data acquired after 6 months

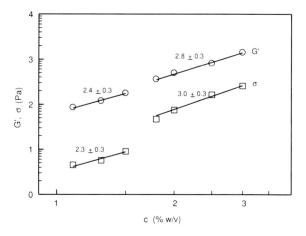

Fig. 3.6 The loss tangent of Laponite dispersions at different concentrations of clay **a** below 2 % (w/v) and **b** above 2 % (w/v). The data was obtained after 6 months

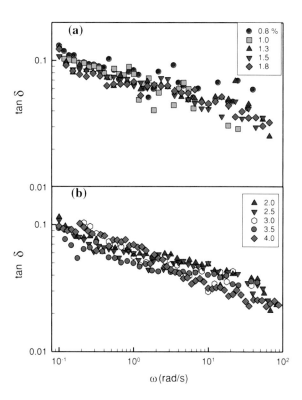

Figure 3.6a shows that the loss tangent is a weak function of the frequency for the concentrations in the range 0.8–1.8 % (w/v), i.e. in sol and gel state of dispersions. However, for $c > 2$ % (w/v), glassy phase, this dependence was stronger as shown in Fig. 3.6b.

3.3 Results and Discussion

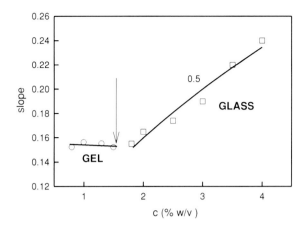

Fig. 3.7 Characteristic slopes of the loss tangent plots are shown as function of clay concentrations, which shows two distinct regions

The slopes of the loss tangents are plotted for all the dispersions of Laponite in Fig. 3.7. Two regions were observed here, in the region of lower concentrations the slopes did not change significantly while above 2 % (w/v) the slope increased as a power-law with exponent 0.5.

The stress-shear rate behavior of the low and high concentration dispersions (gel and glass samples after 6 months) were confirmed from the shear rate dependence data presented in Fig. 3.8. This observation indicated that these colloidal dispersions were non-Newtonian materials. The exponent, α increased from 0.73 to 1 in the gel samples, and remained constant in glass phase [above 1.8 % (w/v)], which is shown in the inset of Fig. 3.8. This conclusion made us look at the growth of yield stress of phase arrested samples (gels and glasses) with aging systematically.

Thus, we have conducted experiments on the evolution of yield stress as function of the aging time which yielded very interesting result and the data is depicted in Fig. 3.9. The yield stress measurements were performed at three clay concentrations and the data was fitted to a logarithmic function described by Eq. 3.3 which gives a universal behavior

$$\sigma = \beta \ln\left(\frac{t_w}{t_m}\right) \tag{3.3}$$

From this experiment it was understood that the domains in the system grew logarithmically with the aging time. These results are consistent with the previous studies made by Rich et al. [28]. This study was extended to examine the evolution of the storage modulus G' with the aging. For this, we have used 3 % (w/v) clay dispersion, the reference to initiation of aging was when the sample was prepared and filtered, which was set as $t_w = 0$. This temporal growth is shown in Fig. 3.9.

For the same sample we performed a pre-shear aging experiment. The aged sample was sheared at a shear rate of 1,000/s for 10 min which destroyed all the self-assembled structures. Thus, the sample was rejuvenated (see the G' data at $t_w = 0$ for original and rejuvenated samples in Fig. 3.9). This sample was

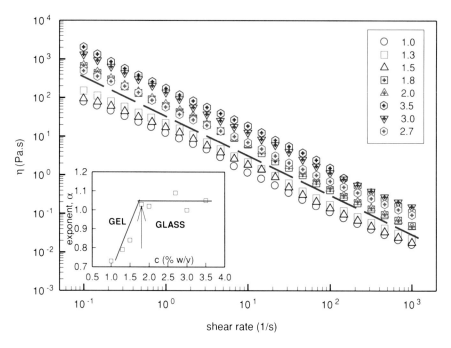

Fig. 3.8 Steady-state viscosity of the system plotted as a function of the shear rate. A power-law with an exponent of α is fitted to the data. Inset shows the variation of α with the concentration, and note that there exist two regions. The dashed line in blue is drawn to indicate different behavior of the Laponite dispersions. The data acquired after 6 months

allowed to age as before under identical conditions. It is interesting to note that the restructuring of the glass has higher elastic modulus at all aging times. The crossing points of G' and G'' were different for these growth processes, and it was 1,200 s for un-sheared sample and 600 s for pre-sheared sample. Now, interestingly, the elastic modulus grew logarithmically given by Eq. 3.4.

$$G' = k \ln\left(\frac{t_w}{t_m}\right) \qquad (3.4)$$

where k is a constant similar to β defined earlier. Note that Eqs. 3.3 and 3.4 depict identical evolution and growth of yield stress and elastic modulus with aging time. The logarithmic growth of yield stress was reported for the Laponite system [28] but the logarithmic growth of elastic modulus was never reported.

3.3.4 Dispersion Homogeneity at $t_w = 0$

In our earlier studies, it was observed that charge and shape anisotropy of clay particles governed the exact phase state of the dispersion [27]. Thus we initiated

3.3 Results and Discussion

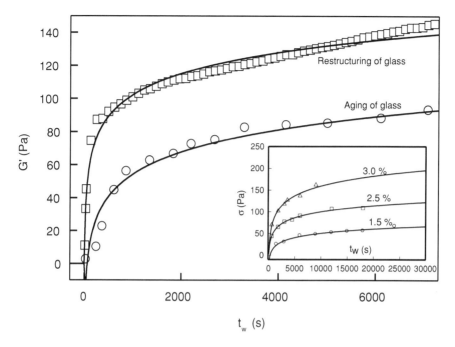

Fig. 3.9 Growth of elastic modulus of the un-sheared (*open circles*) and the pre-sheared (*open squares*) samples of the glass of Laponite (3 % w/v). Inset shows the evolution of yield stress of gel and glass samples with waiting time for 1.5, 2.5, 3 % (w/v) dispersions at room temperature

our investigations by examining the concentration dependent sample homogeneity of clay dispersions. This was best achieved from the dispersive measurement of in-phase storage modulus $G'(\omega)$ and out-of-phase loss modulus $G''(\omega)$ [9, 10, 37]. The correlation between these reveals the sample homogeneity behavior clearly.

The phase homogeneity in polymer solutions and melts is often deduced from Cole–Cole plot where the imaginary part of the complex modulus (G'') is plotted as function of the real part (G') [9]. Typically, in a melt, at very low frequency, viscous behavior is observed whereas at higher frequencies elastic properties dominate. In this formalism, $G^*(\omega) = G'(\omega) + iG'(\omega)$ and the low and high frequency viscosity values are given by G_0 and G_∞ respectively. The Cole–Cole empirical expression is written as

$$G^* - G_\infty = \frac{(G_0 - G_\infty)}{[1 + (j\omega\tau_{cc})^{1-\alpha}]} \; ; \; 0 < \alpha < 1 \qquad (3.5)$$

The aforesaid expression is interpreted as arising from a superposition of several Debye relaxations [9, 10, 37]. The mean relaxation time is given by τ_{cc}, which increases with the concentration of Laponite ($\tau_{cc} = 0.021$ s for 1.5 % (w/v), 0.053 s for 1.8 % (w/v), 0.074 s for 2.0 % (w/v), 0.17 s for 2.5 % (w/v) and 0.23 s for 3 % (w/v)). This plot has been used extensively to map homogeneity of soft matter systems and their composites. For a homogeneous phase, the Cole–Cole

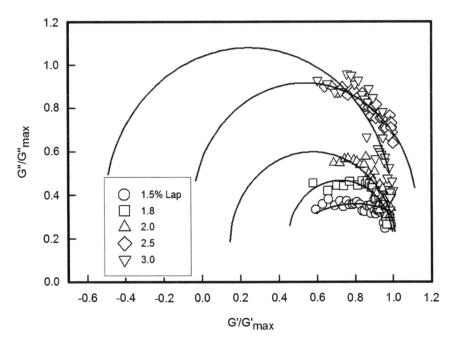

Fig. 3.10 Cole–Cole plot for the specified concentrations of Laponite before aging. The data were normalized by dividing G' with G'_{max} and G'' with G''_{max}. The full line is the theoretical Cole–Cole plot calculated numerically using Eq. 3.5

plot is a perfect semicircle ($\alpha = 0$) with a single relaxation time. The relaxation time of the Laponite samples increase with the concentration of the clay as shown in the Fig. 3.10, but the plots were observed to retain their semicircle feature implying that the nascent dispersions were homogeneous.

3.3.5 Phase Diagram

The hydration data in hand allowed us to construct a 3D-phase diagram of salt-free Laponite dispersion in water for various instances of aging. The 3,200 cm^{-1} Raman band was found to be quite sensitive to clay concentration (Fig. 3.3) and it distinctively demarcated the sol, gel and glassy phases. Note that the scattered light intensity growth plots could not distinguish between sol and gel states (Fig. 3.4). However, the yield stress, storage modulus and loss tangent data could identify the gel and glass regions clearly (Figs. 3.5 and 3.7). All the aforesaid information taken together concluded the following: (i) for $c < 1$ % (w/v) dispersion exists in sol phase (ii) for $1 < c < 2$ % (w/v) networked structures prevailed to yield clay gels and (iii) for solid content higher than 2 % (w/v) disordered random orientation of clay particle gives rise to a glassy phase formation. Similar kind of conclusion

3.3 Results and Discussion

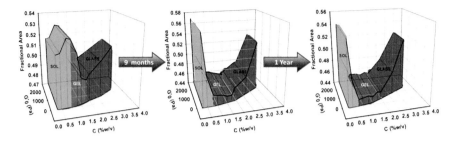

Fig. 3.11 Evolution of three dimensional phase diagram of salt-free dispersion of Laponite at room temperature with aging, without taking into account the phase separation in the sol state

was arrived at by [29], through results obtained from different experiments. Herein, for the first time, the temporal evolution of phase diagram based on hydration is being presented in Fig. 3.11. We have proposed the time-dependent phase diagram without taking into account the phase separation of the sol, which will occur at much longer timescales (at about 3.5 years) than our time of observation (up to 1 year). The aging times depicted in the Fig. 3.11 pertain to data obtained after 9 and 12 months to show the saturation in the gel and glass state.

3.4 Conclusion

Laponite dispersions in water at room temperature in the concentration range 0.3–3.5 % (w/v) was examined by an array of experimental tools over a period of one year. The temporal growth of the self-assembled colloidal structures was probed by light scattering studies which revealed the existence of aggregating structures in both sol and gel states as opposed to the glass phase where the presence of an amorphous arrested phase was indicated. Both in the sol and gel regime ($c < 1.8$ % (w/v)), the intensity of light scattered $I(q, c)$ scaled with concentration c as, $I(q, c) \sim c^\alpha$ with $\alpha = 0.95$ at $t_w = 0$, and 0.63 after $t_w = 6$ months implying that this temporal growth resulted from the formation of colloidal gel whereas in the glass phase ($c \geq 2$ % (w/v)) scattered intensity from samples remained constant ($\alpha = 0$).

The hydration of these anisotropic charged platelets was studied using Raman spectroscopy in a systematic and structured manner over the same period which unambiguously established the presence of three phase states of the dispersion, with each of these aging in its characteristic way. This data was supplemented by the information obtained from rheological mapping of the samples. But, the measured yield stress, storage modulus and loss tangent data could not provide any clear demarcation between the aforesaid phases. However, it has been shown that an aged glass can be made to rejuvenate by subjecting the sample to appropriate shearing. Interestingly, this new born sample follows exactly the same aging trajectory it had followed in its earlier life. This aging process was described by a logarithmic function. A time-dependent 3D-phase diagram has been proposed for the salt-free Laponite dispersion in water which clearly shows different regions of sol, gel and glass. In summary, it is concluded that hydration data determined from Raman spectroscopy could clearly distinguish the phase states and their evolution with aging.

References

1. B. Abou, D. Bonn, J. Meunier, Aging dynamics in a colloidal glass. Phys. Rev. E: Stat. Phys., Plasmas, Fluids, Relat. Interdiscip. Top. **64**, 021510–021516 (2001)
2. Y. Aray, M. Marquez, J. Rodríguez, S. Coll, Y. Simón-Manso, C. Gonzalez, D.A. Weitz, Electrostatics for exploring the nature of water adsorption on the laponite sheets. J. Phys. Chem. B. Surface **107**, 8946–8952 (2003)
3. P. Ball, Water as an active constituent in cell biology. Chem. Rev. **108**, 74–108 (2008)
4. M. Bellour, A. Knaebel, J.L. Harden, F. Lequeux, J.P. Munch, Aging processes and scale dependence in soft glassy colloidal suspensions. Phys. Rev. E **67**, 031405–031408 (2003)
5. B.J. Berne, R. Pecora, Dynamic light scattering with applications to chemistry, biology and physics (Wiley-Interscience, New York, 1976)
6. D. Bonn, S. Tanase, B. Abou, H. Tanaka, J. Meunier, Laponite: aging and shear rejuvenation of a colloidal glass. Phys. Rev. Lett. **89**, 015701 (2002)
7. H.N. Bordallo, L.P. Aldridge, G.J. Churchman, W.P. Gates, M.T.F. Telling, K. Kiefer, P. Fouquet, T. Seydel, S.A.J. Kimber, J. Phys. Chem. C **112**, 13982–13991 (2008)
8. D.M. Carey, G.M. Korenowski, Measurement of the raman spectrum of liquid water. J. Chem. Phys. **108**, 2669–2675 (1998)
9. K.S. Cole, R.H. Cole, Dispersion and absorption in dielectrics I. alternating current characteristics. J. Chem. Phys. **9**, 341–351 (1941)
10. D.W. Davidson, Dielectric relaxation in liquids: I. The representation of relaxation behaviour. Can. J. Chem. **39**, 571–594 (1961)
11. R.R. Desai, J.A.E. Desa, V.K. Aswal, Hydration studies of bentonite clay. AIP Conf. Proc. **197**, 1447–1448 (2012)
12. W.P. Gates, H.N. Bordallo, L.P. Aldridge, T. Seydel, H. Jacobsen, V. Marry, G.J. Churchman, Neutron time-of-flight quantification of water desorption isotherms of montmorillonite. J. Phys. Chem. C **116**, 5558–5570 (2012)
13. S.C. Glotzer, M.J. Solomon, Anisotropy of building blocks and their assembly into complex structures. Nat. Mater. **6**, 557–562 (2007)
14. F.G. Sanchez, F. Jurányi, T. Gimmi, L. Van Loon, T. Unruh, L.W. Diamond, Translational diffusion of water and its dependence on temperature in charged and uncharged clays: a neutron scattering study. J. Chem. Phys. **129**, 174706–174716 (2008)
15. A.Y. Grosberg, T.T. Nguyen, B.I. Shoklovskii, Colloquium: the physics of charge inversion in chemical and biological systems. Rev. Mod. Phys. **74**, 329–345 (2002)
16. J. Israelachvili, H. Wennerstrom, Role of hydration and water structure in biological and colloidal interactions. Nature **379**, 219–225 (1996)
17. S. Jabbari-Farouji, E. Eiser, G.H. Wegdam, D. Bonn, Ageing dynamics of translational and rotational diffusion in a colloidal glass. J. Phys.: Condens. Matter **16**, 471–477 (2004)
18. S. Jabbari-Farouji, G.H. Wegdam, D. Bonn, Gels and glasses in a single system: evidence for an intricate free-energy landscape of glassy materials. Phys. Rev. Lett. **99**, 065701–065704 (2007)
19. A. Knaebel, M. Bellour, J.P. Munch, V. Viasnoff, F. Lequeux, J.L. Harden, Aging behavior of laponite clay particle suspensions. Europhys. Lett. **52**, 73–79 (2000)
20. P. Komadel, J. Hrobáriková, L. Smrčok, B. Koppelhuber-Bitschnau, Hydration of reduced-charge montmorillonite. Clay Miner. **37**, 543–550 (2002)

21. N. Malikova, A. Cadene, V. Marry, E. Dubois, P. Turq, Diffusion of water in clays on the microscopic scale: modeling and experiment. J. Phys. Chem. B **110**, 3206–3214 (2006)
22. V. Marry, E. Dubois, N. Malikovas, S. Durand-Vidal, S. Longevilles, J. Breu, Water dynamics in hectorite clays: infuence of temperature studied by coupling neutron spin echo and molecular dynamics. J. Environ. Sci. Technol. **45**, 2850–2855 (2011)
23. V. Marry, B. Rotenberg, P. Turq, Structure and dynamics of water at a clay surface from molecular dynamics simulation. Phys. Chem. Chem. Phys. **10**, 4802–4813 (2008)
24. S. Morikubo, Y. Sekine, T. Ikeda-Fukazawa, Structure and dynamics of water in mixed solutions including laponite and PEO. J. Chem. Phys. **134**, 044905–044909 (2011)
25. P. Patrice, J.M. Laurent, W. Fabienne, M.F. Anne, D. Alfred, J. Phys. Chem. C **116**, 17682–17697 (2012)
26. F. Pignon, J.M. Piau, A. Magnin, Structure and pertinent length scale of a discotic clay gel. Phys. Rev. Lett. **76**, 4857–4860 (1996)
27. R.K. Pujala, N. Pawar, H.B. Bohidar, Universal sol state behavior and gelation kinetics in mixed clay dispersions. Langmuir **27**, 5193–5203 (2011)
28. J.P. Rich, J. Lammerding, G.H. McKinley, P.S. Doyle, Nonlinear microrheology of an aging, yield stress fluid using magnetic tweezers. Soft Matter **7**, 9933–9943 (2011)
29. B. Ruzicka, E. Zaccarelli, A fresh look at the laponite phase diagram. Soft Matter **7**, 1268–1286 (2011)
30. B. Ruzicka, E. Zaccarelli, L. Zulian, R. Angelini, M. Sztucki, A. Moussaïd, T. Narayanan, F. Sciortino, Observation of empty liquids and equilibrium gels in a colloidal clay. Nat. Mater. **10**, 56–60 (2011)
31. B. Ruzicka, L. Zulian, E. Zaccarelli, R. Angelini, M. Sztucki, A. Moussaïd, G. Ruocco, Competing interactions in arrested states of colloidal clays. Phys. Rev. Lett. **104**, 085701–085704 (2010)
32. B. Ruzicka, L. Zulian, G. Ruocco, More on the phase diagram of laponite. Langmuir **22**, 1106–1111 (2006)
33. T. Seydel, L. Wiegart, F. Juranyi, B. Struth, H. Schober, Unaffected microscopic dynamics of macroscopically arrested water in dilute clay gels. Phys. Rev. E **78**, 061403–061408 (2008)
34. H. Tanaka, S. Jabbari-Farouji, J. Meunier, D. Bonn, Kinetics of ergodic-to-nonergodic transitions in charged colloidal suspensions: aging and gelation. Phys. Rev. E **71**, 021402–021411 (2005)
35. J.T. Trevors, G.H. Pollack, Hypothesis: the origin of life in a hydrogel environment. Prog. Biophys. Mol. Biol. **89**, 1–8 (2005)
36. C.G. Venkatesh, S.A. Rice, J.B. Bates, A raman spectral study of amorphous solid water. J. Chem. Phys. **63**, 1065–1071 (1975)
37. T.C. Warren, J.L. Schrag, J.D. Ferry, Infinite-dilution viscoelastic properties of poly-γ-benzyl-L-glutamate in helicogenic solvents. Biopolymers **12**, 1905–9015 (1973)
38. E. Zaccarelli, Colloidal gels: equilibrium and non-equilibrium routes. J. Phys.: Condens. Matter **19**, 323101–323150 (2007)

Chapter 4
Anisotropic Ordering in Nanoclay Dispersions Induced by Water–Air Interface

Abstract This chapter reports the kinetics of interface induced generation and propagation of arrested phase caused anisotropy in Laponite dispersions probed by depolarized light scattering experiments. Growth of anisotropy as function of waiting time and temperature have been explored in detail. The relaxation dynamics of the dispersion induced a water–air interface has been studied systematically.

4.1 Introduction

Many interesting phenomenon occurring in material science are caused at the interface of different fluids [4, 16, 33], water–air interface being the most common amongst them. Clay dispersions are very interesting soft matter systems that produce various concentration and age dependent arrested phases, which are of both industrial and scientific importance. The non-trivial phase diagrams exhibited by these dispersions owe their existence to the anisotropy associated with both the geometrical shape and charge on these discotic particles.

Colloidal clays have recently emerged as complex model systems with a very rich phase diagram, encompassing fluid, gel and glassy states [1, 3, 5–9, 22, 24, 25–27]. Often these disordered phases are found to interfere with ordered ones, namely nematic and columnar phases. Aging is the most commonly observed feature in disordered states of matter. Glassy materials are not completely frozen, but continue to relax at slower and slower rates as they age towards equilibrium. The phenomenon of aging in clay dispersions is under intense investigation by several research groups [10, 20, 21, 25, 28–31] and it continues to yield interesting features, which are helpful in understanding the complexities observed in biological systems. The effect of temperature is clearly manifested in the properties of soft matter systems. Recently Hansen et al. [11] observed swelling transition in passive clay systems induced by heating. Temperature induced orientational ordering in the mixed clay dispersions has been reported by us in a previous study [23].

The phenomenon of dynamic arrest is clearly observed in clay dispersions which have been extensively probed [1, 6, 20, 21, 25, 26]. Several mechanisms for dynamical arrest have been identified and a rich phenomenology has been predicted both at high re-entrant liquid–glass line, attractive and repulsive glasses and low clay concentrations where gelation was shown to occur from different thermodynamic routes. Recently, the existence of equilibrium gels in systems with intrinsic anisotropy (patchy colloids) has been observed from experimental and numerical studies [27].

Laponite is a synthetic clay that is widely used as a rheology modifier in industrial applications such as paints, varnishes, cosmetics and polymer nanocomposites [12, 32]. To investigate the formation of multiple arrested states, charged colloidal clay made of nanometer-sized discotic platelets, have emerged as suitable candidates. The structural anisotropy associated with these particles, combined with the presence of attractive and repulsive terms in their interaction potentials, makes the phase diagram of such colloidal systems very complex. In particular, Laponite displays a nontrivial aging dynamics replete with multiple arrested states.

There have been many reports on the study of orientational ordering of clays [13, 15, 18, 19, 31]. Recently, the phenomenon of the interface induced ordering was observed by Shahin et al. [31] in Laponite suspensions using optical birefringence studies. An important question arises here: how does the interface effect the particle dynamics and its ordering? Herein this Chapter, we address this issue through a series of controlled experiments performed on aging laponite dispersions. We report on the ordering of laponite platelets as the dispersion developed optical anisotropy starting at the water–air interface. This propagated into the bulk over many orders of magnitude compared to the platelet size with aging time. We have used depolarized laser light scattering to investigate this ordering process quantitatively and further studied the relaxation dynamics from dynamic structure factor measurements. The effect of insulation of the water–air interface, temperature and aging time on the aforesaid phenomenon was also studied systematically.

4.2 Experimental Geometry

The experiment required that scattered light was to be collected from different sample depths with respect to the water–air interface. Since, in a standard goniometer geometry attached to a light scattering spectrometer it is not possible to collect scattered light emanating from various sample depths, we had to make a precision sample holder and adapt it to the sample compartment of the goniometer.

It contained the micrometer screw jack mounted on a disc shaped structure as shown in Fig. 4.1. The screw jack was connected to a cylindrical neck below the circular disc. The top of the cylindrical glass cell which contained the sample could be inserted into this neck and fastened with three symmetrically placed teflon screws. This placed the sample cell firmly inside the scattering chamber of the

4.2 Experimental Geometry

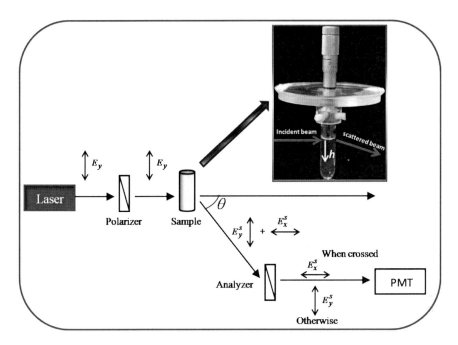

Fig. 4.1 Schematic illustration of the set up used to study the depth dependent growth of anisotropy in the system. The sample holder, micrometer screw jack and the sample are shown in the picture

goniometer. The jack was calibrated with the main scales and 25 circular scales. This meant that the jack was capable of producing rotational as well as the translational displacement. By moving the jack in clockwise direction we could move the sample downwards whereas its anticlockwise motion lifted the sample in upward direction. So, the rotational as well as the translational displacement of the sample was possible that was required for our inspection as we wanted to observe the anisotropy as function of depth from the interface to bulk inside the sample. This arrangement facilitated the vertical displacement of the cell with the precision of 0.255 µm which was the least count of the jack. Thus, the displacements caused during the measurements were extremely precise.

The analyzer was kept at right angle to the direction of the propagation of the laser light so as to minimize the effect of the stray light and the presence of an interference filter ensured that signal to noise ratio was robust. The angle in the analyzer was adjusted either to 0° or 90° respectively as per the requirement. The 0° alignment of the analyzer meant that only the parallel component (I_{VV}) of the scattered light passed through it to the detector whereas the 90° alignment of the analyzer ensured passage of only perpendicular component (I_{VH}) of the scattered light. A built-in temperature controller was used to control the temperature during the experiments which was fixed to 25 °C in our case.

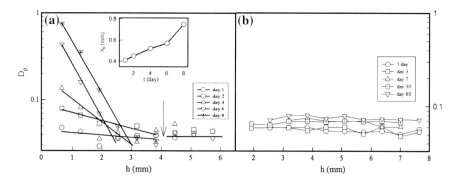

Fig. 4.2 Variation of depolarization ratio, D_p as function of depth from water–air interface of the sample at different aging times. *Left* (**a**) and *right* (**b**) figures depict data for 3 and 1.5 % (w/v) dispersions respectively. The data in **a** was least-squares fitted to the exponential decay given by Eq. 4.4 (with $X^2 = 0.99$), and the decay lengths are plotted as function of waiting time shown in *inset*. The *arrow* divides the exponential fitting and the constant values. The size of the *symbol* represents the error in the measurement

4.3 Experimental Results

4.3.1 Anisotropic Ordering at the Water–Air Interface

In this work, we have reported on Laponite suspensions that developed optical anisotropy over many days while exhibiting transformation into a completely jammed state. The geometrical anisotropy associated with the scattering particles depolarizes the incident light, and the scattered electrical field can be decomposed into the parallel $E_{VV}(q, t)$, and perpendicular $E_{VH}(q, t)$ components, with respect to the direction of the incident polarization. These quantities fluctuate due to the random translational and rotational motion of the particles, and one can define two distinct dynamic structure factors or electric field correlation functions using these components of the scattered field.

Interface induced anisotropy was developed at the liquid–air interface whereas it was absent in the liquid–liquid interface which will be discussed later. Growth in optical anisotropy was monitored systematically with aging time and depth of the sample. Initially, we had used 3 % (w/v) of Laponite to study the anisotropic growth. Figure 4.2 illustrates a nice evolution of depolarization ratio with depth. Similar observation was made by Shahin et al. [31], where they have used optical birefringence measurements to find the interface induced ordering. The solid concentration of 3 % (w/v) belongs to the glassy state of the Laponite system. The system in glass phase not only undergoes aging but develops spatial ordering which starts at the interface between liquid and air, and percolates into the bulk which is clearly seen from Fig. 4.2. The depolarization ratio which defines the anisotropic ordering of the system was found to grow with waiting time.

4.3 Experimental Results

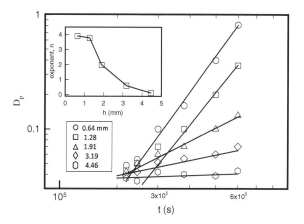

Fig. 4.3 Time dependent evolution of D_p at various depths of 3 % (w/v) Laponite dispersion. This data was least-squares fitted to the power-law relations given by Eq. 4.3. *Inset* shows the variation of power-law exponent with depth from interface

Depolarization ratio was defined as follows [2]

$$D_p = \frac{I_{VH}}{I_{iso}} \qquad (4.1)$$

and

$$I_{iso} = I_{VV} - \frac{4}{3}I_{VH} \qquad (4.2)$$

where I_{VH} and I_{VV} are the depolarized and polarized components of the scattered intensity.

Initially the growth in D_p was slow while at long time scales it was rapid. In Fig. 4.2, D_p is shown as function of the depth of the sample. It is evident from this plot that there was substantial development of anisotropy at the interface which progressively propagated into the bulk and it vanished slowly beyond the depth of 5 mm. The ordering that was observed in this system was irreversible with aging and depth. Since 3 % (w/v) of Laponite is assumed to be in the glass state, it could be melted by the dilution test. The D_p versus aging time t_w data shown in Fig. 4.3 could be least-squares fitted to a power-law function for a given depth h from the interface given by

$$D_p \sim t_w^n \qquad (4.3)$$

The power-law exponent n increased from 0.1 to 4.0 as one moved away from the interface into the bulk. Similarly, a strong dependence of D_p on depth h was noticed and an exponential decay behavior could be established. This is given by

$$D_p \sim \exp(-h/h_0) \qquad (4.4)$$

Here the value of the decay length h_0, increased from 0.4 to 0.75 mm with aging time (see Fig. 4.2a).

Interestingly, there was no development of anisotropy observed at lower concentrations of Laponite. We performed measurements up to 60 days but did not observe any significant change in the value for D_p. A typical depiction of this is illustrated in Fig. 4.2b.

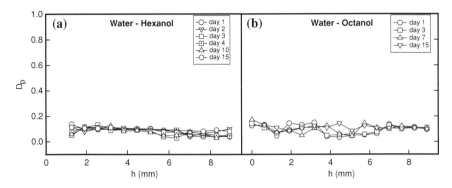

Fig. 4.4 Effect of covering the surface of the Laponite dispersion with hydrophobic solvent layers. D_p is shown as function of depth at different aging times. **a** Water–hexanol and **b** water–octanol interface

4.3.2 Effect of Water–Hydrophobic Liquid Interface

A very interesting observation was made during our investigations which clearly illustrated the role played by the interface in causing anisotropic ordering of the platelets. In this study, we spread a thin layer water immiscible hydrophobic solvent hexanol (also octanol) on the surface of the dispersion, thereby changed the water–air interface to water–hexanol interface. Surprisingly, no growth in anisotropy was observed in both the cases even at the longer times, though the samples entered the non-equilibrium state (glass). This observation is similar to Shahin et al. [31] for the water–oil interface. This result is depicted in Fig. 4.4 pertaining to the two cases where hexanol and octanol layers present on the Laponite dispersion provided the water–hexanol and water–octanol interface respectively.

4.3.3 Effect of Temperature

It was felt imperative to examine if temperature had any role to play in the ordering of the platelets. There was no enhancement or the improvement of ordering in the hydrophobic liquid covered surfaces. For the samples having water–air interface the effect of temperature induced ordering was observed. This was similar to the one we had observed in our previous study on mixed clay dispersions where the temperature induced ordering in the bulk of the sample was observed [22, 23]. Figure 4.5 describes the effect of temperature on sample anisotropy for different depths and Fig 4.5 is similar to the observation made by Shahin et al. [31]. It is clearly seen at higher temperature.

The depolarization ratio was observed to decay exponentially as shown in Eq. (4.4), where h_0 represents the typical decay length. The value of h_0 thus

4.3 Experimental Results

Fig. 4.5 Depolarization ratio shown as function of depth of the sample at different temperatures as indicated against each *curve*

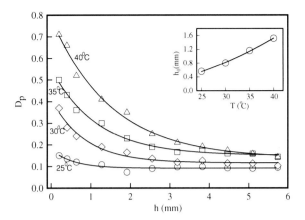

obtained was found to increased with temperature as shown in the inset of Fig. 4.5, and could follow a power-law behavior given by

$$h_0 \sim T^\gamma \tag{4.5}$$

where $\gamma = 2.0 \pm 0.2$.

4.3.4 Relaxation Dynamics

The relaxation dynamics of the system was studied using dynamic light scattering measurements. We have analyzed the dynamic structure factor data both in the polarized and depolarized mode. The dynamic structure factors thus obtained were plotted as shown in Fig. 4.6 and analyzed using the same protocol that is described in our previous work [20]. We have taken the heterodyne contribution into account while dealing with these non-ergodic samples.

The dynamic structure factor $g_1(q, t)$ could be described as

$$g_1(q, t) = a \exp\left(-\frac{t}{\tau_1}\right) + (1-a) \exp\left(-\frac{t}{\tau_2}\right)^\beta \tag{4.6}$$

where a and $(1 - a)$ are the weights of the two contributions τ_1 and τ_2, which in turn are the fast and the slow mode relaxation times respectively, and β is the stretching parameter. The fast mode relaxation time τ_1 is related to the inverse of the short-time diffusion coefficient D_s as $\tau_1 = 1/D_s q^2$. The presence of stretched exponential function has been reported in the past for clay dispersions, since it has been found empirically that it provides effective description of the slow mode relaxation processes found in arrested systems [9, 20, 26]. Equation 4.6 gave a good description to both the polarized and the depolarized dynamic structure factors $g_1^{VV}(q, t)$ and $g_1^{VH}(q, t)$.

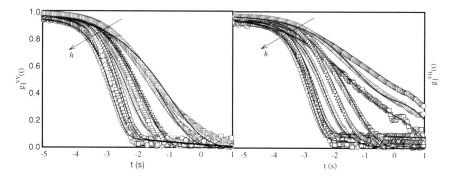

Fig. 4.6 Evolution of the dynamic structure factors, $g_1^{VV}(q,t)$ (*left panel*) and $g_1^{VH}(q,t)$ (*right panel*) at different depths at 25 °C. The *arrows* indicate increasing depths starting from 0.2 to 5 mm. *Solid lines* represent the fitting curve to Eq. 4.6

Fig. 4.7 Characteristic slow mode relaxation time plotted as function of depth for polarized and depolarized components at different temperatures as indicated. The exponential decay lengths are indicated against each *curve*

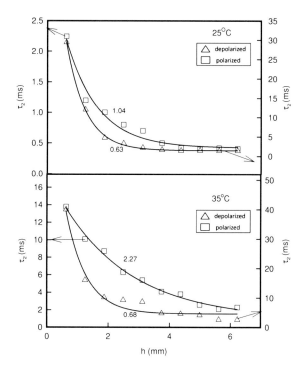

The slow mode relaxation time of both the components are compared in Fig. 4.7 for results obtained at different temperatures. This relaxation time τ_2 was observed to freeze early at the interface compared to same in the bulk because the self-arrested structures relax slowly close to the water–air interface. It is interesting to note that the relaxation time of polarized and depolarized components behaved differently. The slow mode relaxation time τ_2 of $g_1^{VV}(q,t)$ was lower

4.3 Experimental Results

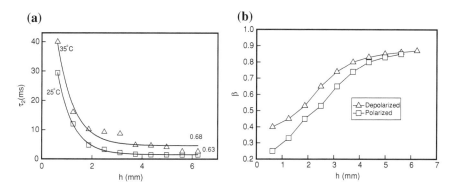

Fig. 4.8 **a** The characteristic slow mode relaxation time with the depth of the sample at two different temperatures 25 and 35 °C in the depolarized mode dynamic structure factor, $g_1^{VH}(q, t)$. The exponential decay lengths are indicated against each curve. **b** Depth dependent evolution of β for polarized and depolarized modes at 25 °C

than that of the other component $g_1^{VH}(q, t)$. The τ_2 values obtained were constant beyond the depth 4 mm. The influence of temperature on this system was also studied and the data is shown in Fig. 4.7. Measurements were done at 25 and 35 °C, which indicated that at higher temperature the system reached the arrested phase quickly. The τ_2 obtained from $g_1^{VH}(q, t)$ data was compiled and the same are compared for different temperatures which is shown in Fig. 4.7.

The slow mode relaxation in both the polarized and depolarized modes was found to decays exponentially with depth of the sample, given by

$$\tau_2 \sim \exp(-h/h_0) \tag{4.7}$$

where h_0 represents the decay length. The parameter h_0 increased with the temperature as shown in Fig. 4.8a. The stretching parameter β increased as we go from interface into bulk as shown in Fig. 4.8b, which implied that the part of interface had reached the arrested state much before the same occurring in the bulk.

It reveals that even after slowing down of the dynamics, the particles still have some freedom to move and reorganize themselves in the process of attaining equilibrium. To realize any possible correlation existing between the ordering, and the relaxation times we have plotted D_p as function of τ_2 at particular depths. We found a scaling behavior could be found between D_p and τ_2 given as

$$D_p \sim \tau_2^\alpha \tag{4.8}$$

Temperature dependence α *is* shown in Fig. 4.9 for polarized depolarized components. Such a scaling relation was never reported in the literature before. The value of α was found to increase with temperature in VH and the increase is negligible in VV case.

Interestingly dilution destroyed the ordering from the interface as shown in Fig. 4.10, which means the sample exhibit ordering even in the glass.

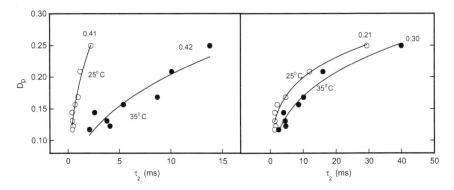

Fig. 4.9 Depolarization ratio plotted as the function of slow mode relaxation time determined from polarized mode dynamic structure factor, $g_1^{VV}(q,t)$ (*left panel*) and depolarized mode dynamic structure factor, $g_1^{VH}(q,t)$ (*right panel*). The power-law exponents are indicated against each curve

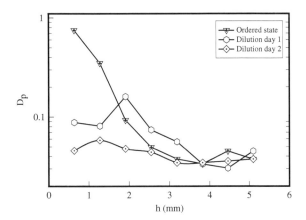

Fig. 4.10 Dilution of the 3 % (w/v) Laponite at different times as indicated. Dilution destroys the oriental ordering which is clear from the graph, where D_p decreased by the addition of water to the sample

4.4 Discussion

We have looked into the kinetics of orientational ordering in Laponite suspensions under various conditions. The study indicates that the ordering wais influenced by mainly four factors: (1) water–air interface, (2) liquid–liquid (water–hydrophobic solvents) interface, (3) waiting time and (4) temperature.

From the relaxation dynamics it was observed that the particles at the water–air interface were arrested earlier than the particles in the bulk. The relaxation time of depolarized component decayed faster than the polarized component. Thus, we could clearly observe the presence of heterogeneity in the aging sample varying with the depth measured from the interface. Mamane et al. [14] made a very interesting observation on the surface fluctuations of an aging Laponite suspension. They showed that as the fluid aged, the dynamics became heterogeneous and those

4.4 Discussion

quakes at the surface were uncorrelated. They used the reflection of a laser beam on the free surface of the clay dispersion, the position of which was sensitive, at the first order, to the local slope of the surface. It was shown that, at large waiting times, the thermally induced surface fluctuations exhibited bursts of activity, and were characterized by non-Gaussian dynamics. They found that the corresponding events induced large changes in the local slope of the free surface at a scale larger than the 60 μm of the beam size and were uncorrelated. In that process particles tried to attain the minimum free energy configuration by arranging themselves with its major axis in horizontal position.

Our measurements found the existence of correlation between the ordering and the relaxation times with aging and with depth of the sample from the interface. Moreover dilution destroyed the ordering implying the presence of strong repulsive interaction between the platelets. Thus this phase could be identified as orientationally ordered glass of colloidal platelets.

But when the surface of the dispersion was covered with a thin layer of hydrophobic solvent, the growth anisotropic ordering was not observed. Even temperature was not able to drive the system to cause the anisotropic ordering. The microstructure of liquid–liquid interface and its interesting dynamics is still exploring. The thickness of the liquid–liquid (water–alcohol) interface is considered to be only a few nanometers [8, 17]. These interfaces are flexible and cannot be fixed at a spatial position. For the water–air interface measurements one may be tempted to understand that the interface is responsible for the ergodicity breaking, but it may not be. The water–alcohol interface is not entirely resisting the system going into the arrested state. It could be possible that the octanol–water interface is acting like a reflecting boundary not allowing the particles to order, stay or stick at the interface for longer times.

4.5 Conclusion

We have studied the relaxation dynamics of interface dependent ordering phenomenon systematically using the light scattering experiments. The geometry that we have used in our experiments conclusively mapped the dynamics of anisotropy growth occurring at the interface and propagating into the bulk with aging. We believe that the water–air interface is responsible for causing this ordering. The density fluctuations at the interface of water–air are much stronger compared to the bulk, which may be responsible for the observed occurrence of anisotropy. The parameters that enhanced the ordering were the aging time and temperature. Higher the temperature higher was the amount of depolarization, and the particles encountered arrested state early. Interestingly, we noticed the presence of heterogeneous relaxations in these dispersions that varied with the depth which was not seen hitherto. When the interface was covered with oil or water immiscible liquids, the surface fluctuations vanished and the dynamics of aging exhibited homogeneous features. Interestingly covering of the surface has significant effect on the arrested dynamics. The ergodicity breaking time is delayed when the surface is covered with water miscible liquids, and further studies in this direction are underway. We infer that pH and salt dependence have to be studied using this set up to gain further insight into the ordering phenomenon though it was studied earlier [31].

References

1. B. Abou, F. Gallet, Probing a nonequilibrium einstein relation in an aging colloidal glass. Phys. Rev. Lett. **93**, 160603 (2004)
2. B.J. Berne, R. Pecora, *Dynamic light scattering with applications to chemistry, biology, and physics* (Wiley Interscience, New York, 1976)
3. D. Bonn, H. Tanaka, G. Wegdam, H. Kellay, J. Meunier, Aging of a colloidal Wigner glass. Europhys. Lett. **45**, 52 (1998)
4. H.K. Choi, M.H. Kim, S.H. Im, O.O. Park, Fabrication of ordered nanostructured arrays using poly (dimethylsiloxane) replica molds based on three-dimensional colloidal crystals. Adv. Funct. Mater. **19**, 1594–1600 (2009)
5. P. Coussot, Q.D. Nguyen, H.T. Huynh, D. Bonn, Coexistence of liquid and solid phases in flowing soft-glassy materials. Phys. Rev. Lett. **88**, 218301 (2002)
6. H.Z. Cummins, Liquid, glass, gel: The phases of colloidal laponite. J. Non-Cryst. Solids **253**, 3891–3905 (2007)
7. M. Dijkstra, J.P. Hansen, P.A. Madden, Gelation of a clay colloid suspension. Phys. Rev. Lett. **75**, 2236 (1995)
8. J. Gao, W.L. Jorgensen, J. Phys. Chem. **92**, 5813 (1988)
9. S. Jabbari-Farouji, G.H. Wegdam, D. Bonn, Ageing dynamics of translational and rotational diffusion in a colloidal glass. J. Phys. Condens. Matter **16**, 471 (2004)
10. S. Jabbari-Farouji, G.H. Wegdam, D. Bonn, Aging of rotational diffusion in colloidal gels and glasses. Phys. Rev. E **86**, 041401 (2012)
11. E.L. Hansen, H. Hemmen, D.M. Fonseca, C. Coutant, K.D. Knudsen, T.S. Plivelic, D. Bonn, J.O. Fossum, Swelling transition of a clay induced by heating. Scientific Reports **2**, 618 (2012)
12. Laponite Technical Bulletin, Laponite Industries Limited, LI04/90/A (1990)
13. B.J. Lemaire, P. Panine, J.C.P. Gabriel, P. Davidson, The measurement by SAXS of the nematic order parameter of laponite gels. Europhys. Lett. **59**, 55 (2002)
14. A. Mamane, C. Fretigny, F. Lequeux, L. Talini, Surface fluctuations of an aging colloidal suspension: evidence for intermittent quakes. Europhys. Lett. **88**, 58002 (2009)
15. C. Martin, F. Pignon, A. Magnin, M. Meireles, V. LeliÃvre, P. Lindner, B. Cabane, Osmotic Compression and Expansion of Highly Ordered Clay Dispersions. Langmuir **22**, 4065 (2006)
16. E.C. Mbamala, H.H. von Grünberg, Charged colloids and proteins at an air–water interface: the effect of dielectric substrates on interaction and phase behavior. Phys. Rev. E **67**, 031608 (2003)
17. D.M. Mitrinovic, Z. Zhang, S.M. Williams, Z. Huang, M.L. Schlossman, J. Phys. Chem. B **103**, 1779 (1999)
18. A. Mourchid, A. Delville, J. Lambard, E. LeColier, P. Levitz, Phase diagram of colloidal dispersions of anisotropic charged particles: Equilibrium properties, structure, and rheology of laponite suspensions. Langmuir **11**, 1942 (1995)
19. F. Pignon, M. Abyan, C. David, A. Magnin, M. Sztucki, In situ characterization by SAXS of concentration polarization layers during cross-flow ultrafiltration of laponite dispersions. Langmuir **28**, 1083 (2012)
20. R.K. Pujala, H.B. Bohidar, Ergodicity breaking and aging dynamics in laponite-montmorillonite mixed clay dispersions. Soft Matter **8**, 6120 (2012)
21. R.K. Pujala, H.B. Bohidar, Slow dynamics, hydration and heterogeneity in laponite dispersions. Soft Matter **9**, 2003 (2013)

References

22. R.K. Pujala, N. Pawar, H.B. Bohidar, Universal sol state behavior and gelation kinetics in mixed clay dispersions. Langmuir **27**, 5193 (2011)
23. R.K. Pujala, N. Pawar, H.B. Bohidar, Landau theory description of observed isotropic to anisotropic phase transition in mixed clay gels. J. Chem. Phys **134**, 194904 (2011)
24. B. Ruzicka, E. Zaccarelli, A fresh look at the laponite phase diagram. Soft Matter **7**, 1268 (2011)
25. B. Ruzicka, L. Zulian, G. Ruocco, Routes to gelation in a clay suspension. Phys. Rev. Lett. **93**, 258301 (2004)
26. B. Ruzicka, L. Zulian, G. Ruocco, More on the phase diagram of laponite. Langmuir **22**, 1106 (2006)
27. B. Ruzicka, E. Zaccarelli, L. Zulian, R. Angelini, M. Sztucki, A. Moussaïd, T. Narayanan, F. Sciortino, Observation of empty liquids and equilibrium gels in a colloidal clay. Nat. Mater. **10**, 56 (2011)
28. A. Shahin, Y.M. Joshi, Irreversible aging dynamics and generic phase behavior of aqueous suspensions of laponite. Langmuir **26**, 4219–4225 (2010)
29. A. Shahin, Y.M. Joshi, Prediction of long and short time rheological behavior in soft glassy materials. Phys. Rev. Lett. **106**, 038302 (2011)
30. A. Shahin, Y.M. Joshi, Hyper-aging dynamics of nanoclay suspension. Langmuir **28**, 5826–5833 (2012)
31. A. Shahin, Y.M. Joshi, S.A. Ramakrishna, Interface-induced anisotropy and the nematic glass/gel state in jammed aqueous laponite suspensions. Langmuir **27**, 14045–14052 (2011)
32. H. Van Olphen, *An Introduction to Clay Colloid Chemistry* (Willey and Sons, New York, 1997)
33. H.H. Wickman, J.N. Korley, Colloidal crystal self-organization and dynamics at the air/water interface. Nature **393**, 445 (1998)

Chapter 5
Phase Diagram of Aging Montmorillonite Dispersions

Abstract This chapter investigates the phase diagram of extensively aged MMT dispersions. Formation of gel through different routes is reported. Distinctive phase separation, equilibrium fluid and equilibrium gels in the t_w-c phase space are observed for the first time ever in MMT dispersions (Fig. 5.1).

5.1 Introduction

Phase stability of dispersions of anisotropic particles has been of much scientific debate in the recent times. Many phase diagrams have been conceived and proposed for colloidal particles having different degree of geometrical anisotropy, namely rods, platelets, disks etc. in their dispersion states [2, 28, 29]. Spherical colloidal particles have been studied extensively in the literature. Clays are discotic platelets with varying aspect ratio and surface charge density. Because of this clay dispersions have exhibited non-trivial and rich phase diagrams, which have attracted much attention in recent years. Suspensions of clays have unique properties, including the ability to form arrested states like gels, glasses and liquid crystalline structures under ambient conditions as function of aging time and solid concentration. Remarkably, compared to spherical colloids, clay platelets exhibit gel and glass phases at very low concentrations which owe their origin to their structure and surface charge heterogeneity.

Montmorillonite (MMT) is one of the natural clays, which has high aspect ratio and is a macroscopically swelling, 'active' clay that has the capacity for taking up large amounts of water to form stable gels. The phase diagram of MMT is least explored in the literature, though some attempts in that direction have been made in the past Michot et al. [16]. Michot et al. [16] studied the phase diagram of MMT using different aspect ratio samples. Recently, we have reported the sol and gel transition behavior of laponite-MMT mixed clay dispersions [21]. On the other hand, visco-elastic properties of MMT suspensions have extensively reported [4, 8, 17, 22, 25, 26]. For instance, rheology behaviour of bentonite slurries as a function of pH, molar ratio of Na^+/Ca^{+2} and with a range of additives such as pyrophosphate, polyphosphate [31]

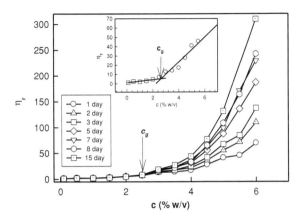

Fig. 5.1 Relative viscosity as function of concentration of clay at different aging times as indicated. The typical threshold concentration c_g is indentified in the inset. *Solid lines* are guide to eye

and SDS surfactant has been reported. According to Lagaly et al. bentonite clay particles form house of cards structures via face (−)/edge (+) (FE) attraction in acidic medium and band-like structures are formed via cation-mediated face (−)/face (−) (FF) attraction in alkaline medium [14]. The strength and nature of these microstructures are shown to be affected by the nature and structure of the adsorbed additives [18]. At relatively low pH, i.e. when the edges are supposedly having a positive charge, the EF configuration would be favored, whereas higher concentration and higher pH would favor the EE configuration. Finally, high charge density on the edges would tend to favor the formation of FF-like structures. Recently, a detailed study on the tactoid formation in Ca Montmorillonite was carried out by [24]. Montmorillonite, natural swelling clay that absorbs water and swells considerably forms a yield stress gel at very low solid concentration, below 4 % (w/v). Laponite, a synthetic discotic clay with a diameter of 25 and 1 nm thickness, also displayed gel-like behavior at an even lower solid concentration of 1 % (w/v) [20, 23]. All the clays are capable of adsorbing water onto their surface. The presence of adsorbed water covering the clay particles produces characteristic cohesive plastic behavior of clay minerals. Recent studies described clay as patchy colloid with limited valence, have the ability to form empty liquids and equilibrium gels [15, 23]. However, considerable confusion persists as far as the aging phase diagram of MMT is concerned. For instance, study of the charge characteristics of swelling clays show that at pH 8 the edge faces of MMT are likely to be negatively charged [14].

In this chapter, we have studied the phase diagram of MMT in salt-free suspensions in normal pH conditions spread over a period of 3.5 years and have noticed that these suspensions undergo nontrivial evolution and aging dynamics.

5.2 Sample Preparation

Na MMT, a hydrated aluminium silicate, was purchased from Southern Clay Products, U.S. and used as received. Chemically it is hydrated sodium calcium aluminium magnesium silicate hydroxide (Na, Ca)$_{0.33}$(Al, Mg)$_2$(SiB$_{4B}$O$_{10}$)(OH)$_2$·nHB$_2$O.

5.2 Sample Preparation

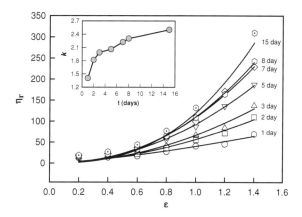

Fig. 5.2 Relative viscosity as function of scaled concentration (ε) as defined in Eq. 5.1, at different aging times. Inset shows the change in the value of power-law exponent with aging time. *Solid lines* are least-square fitting of data to Eq. 5.1

An aqueous dispersion of MMT was prepared by dispersing it in deionized water at pH 8.5. The sample was stirred vigorously using magnetic stirrer for 48 hours to ensure complete dissolution of the clay particles and the fractionation was done as described in [6]. We have carried out experiments in the broad concentration range 0.1–6 % (w/v). All the experiments were performed under room temperature conditions, temperature 25 °C and relative humidity <40 %.

5.3 Results and Discussion

5.3.1 Time-Dependent Viscosity

Time-dependent zero-shear viscosity measurements were performed to establish the growth kinetics. Viscosity of the clay dispersions was found to increase with the concentration of solid content clearly exhibiting aging effects. The onset of aging occurred beyond a certain clay concentration, called the threshold concentration, $c_g = 2.5$ % (w/v), was determined from change in slope of the relative viscosity versus concentration profile data which is shown in Fig. 5.1. The viscosity data was least-squares fitted to Eq. 5.1, where the viscosity grows as function of concentration as $c \to c_g$ (Fig. 5.2). The power-law exponents in the scaled quantities were found to increase with the waiting time t_w, which clearly indicated the evolution of network structures inside the dispersions.

The relative viscosity with the scaled concentration above the threshold concentration is plotted in Fig. 5.2 which could be described by the power-law function given by

$$\eta_r \sim \varepsilon^k; \quad c > c_g \tag{5.1}$$

where $\eta_r = \eta/\eta_0$, η is the viscosity of the dispersion and η_0 is solvent viscosity. The clay concentration is c and its gelation concentration is c_g and $\varepsilon = \left|\left(c/c_g\right) - 1\right|$.

The exponent, k increased from 1.4 to 2.5 with the aging time. It is clear from Figs. 5.1 and 5.2 that the self-assembly of the platelets show distinctive aging effect beyond c_g. The importance and significance of the observed power-law exponent k will be discussed later.

5.3.2 Visco-Elasticity

Time-dependent rheology studies were performed on the dispersions in the frequency sweep mode to probe the visco-elastic attributes of the samples. According to Maxwell model, one can express the in-phase modulus G' and out of phase modulus G'' as function of frequency as

$$G' = \frac{G_0 \omega^2 \tau^2}{1 + \omega^2 \tau^2} \quad \text{and} \quad G'' = \frac{G_0 \omega \tau}{1 + \omega^2 \tau^2} \tag{5.2}$$

where G_0 is the plateau elastic modulus, τ is the relaxation time, and ω is the angular frequency. In the limiting case $\omega \to 0$, the relations will become $G' \sim \omega^2$ and $G'' \sim \omega$ which are considered signatures of the Maxwellian model. Data indicated that for the samples with $c < c_g$ Maxwellian behavior was observed and effect of aging was absent. It appeared that freshly prepared dispersions exhibited Maxwellian behavior. As the samples aged this attribute was lost. In comparison, deviations from Maxwellian behavior was noticed for the samples with $c > c_g$, and with aging this phenomena became more profound which is clearly seen from the Figs. 5.3 and 5.4.

Eventually the viscous and elastic modulii of the dispersions increased as the material started revealing non-Newtonian behavior. The system slowly gets arrested in time as is the case with the other clay systems [20, 23]. The elastic modulus of the sample showed a plateau invariant of the frequency and it behaved as a solid-like material. Higher the solid content lesser was the time required to reach the solid-like phase which is clearly shown in Figs. 5.3 and 5.4.

The low frequency elastic modulus data could be described by the power-law behavior function given by

$$G' \sim \varepsilon^t; \quad c > c_g \tag{5.3}$$

The measurements were done with angular frequency is 1 rad/s at a constant stress of 0.6 Pa. The value of the exponent varies with waiting time as shown in Fig. 5.5. We will use the exponents to understand the percolation model of the system in a separate section.

5.3.3 Steady State Viscosity and Yield Stress

The flow behavior of any system is characterized from the relationship between the shear stress and the shear rate. The shear rate is defined as the change of shear strain per unit time, and the shear stress as the tangential force applied per unit area.

5.3 Results and Discussion

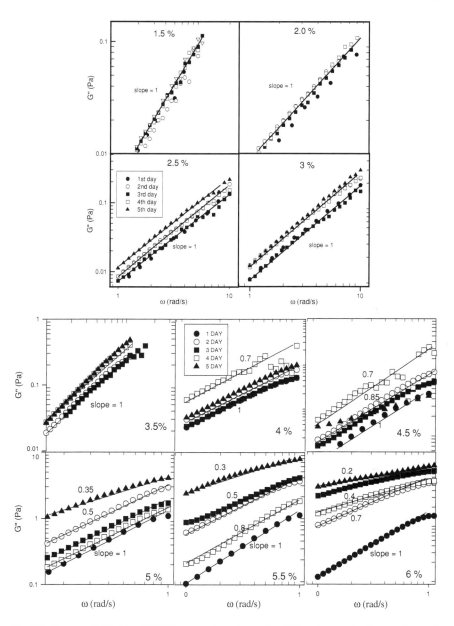

Fig. 5.3 Loss modulii data of MMT suspensions shown for different concentrations and at various aging times as indicated. Note that the deviation from the Maxwellian behavior starts when $c > c_g$ ($= 2.5$ % w/v)

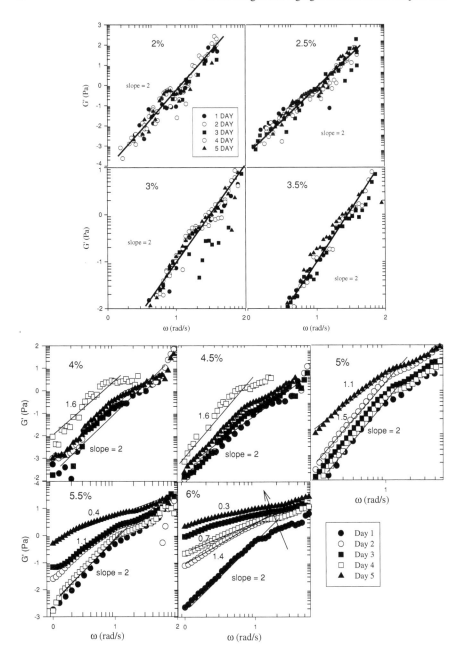

Fig. 5.4 Storage modulii data of MMT suspensions is shown different for concentration and at various aging times as indicated. Note that the deviation from the Maxwellian behavior starts when $c > c_g$ (2.5 % (w/v)) as the exponent started decreasing with aging which is indicated against each curve

5.3 Results and Discussion

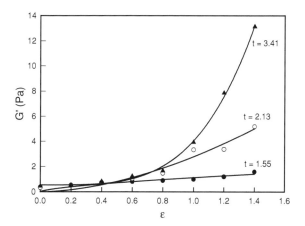

Fig. 5.5 Evolution of storage modulus as a function of scaled concentration at different aging times [1 day (*dark circles*); 2 day (*open circles*); 5 day (*dark triangles*)]. The exponents obtained from Eq. 5.3 were indicated against each curve

The ratio of shear stress to shear rate is called apparent viscosity which is a measure of resistance to the flow of the fluid under consideration. This is represented as

$$\eta_a = \tau/\dot{\gamma} \qquad (5.3)$$

where τ is the shear stress and $\dot{\gamma}$ represents the shear rate. The investigated clay dispersions exhibited an apparent viscosity η_a that decreased with increasing shear rate up to $1{,}000\ \mathrm{s}^{-1}$ due to the rupture of the self-assembled microstructures present in the dispersion. The higher viscosity of MMT dispersion is a consequence of stronger electrostatic interactions prevailing between the platelets. The flow curves for the studied dispersions are shown in Fig. 5.6 for $c > c_g$, the data shows pseudoplasticity (shear thinning) having a yield stress defined as the stress above which the material flows like a viscous fluid. This observation indicated that these colloidal dispersions were non-Newtonian materials. The exponent, m obtained from Eq. 5.4 increased from 0.83 to 1 in the samples which is shown in the inset of Fig. 5.6. This observation made us look at the growth of yield stress of phase arrested samples.

The viscosity of the shear thinning fluid could be described by power-law function given by

$$\eta_a \sim \dot{\gamma}^{-m} \qquad (5.4)$$

The observed stress response of the samples upon aging indicated about the emergence of a well defined yield stress. In order to quantify this, measured flow curves were analyzed with the Herschel–Bulkley formalism given by [12]

$$\tau(\dot{\gamma}) = \tau_0 + K\dot{\gamma}^n \qquad (5.5)$$

where τ_0 is the yield stress, K represents the consistency factor and n is a parameter that characterizes the pseudoplasticity of the system. Generally, the yield stress τ_0 is determined as the intercept of the flow curve at zero shear rate (Fig. 5.7). The value thus obtained can be strongly influenced by the shear rate range used and by

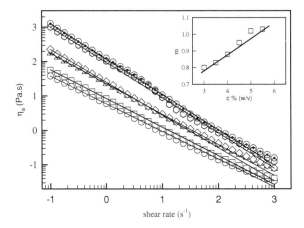

Fig. 5.6 Steady-state viscosity of the system plotted as a function of the shear rate for different clay concentrations (*open circle*) 3.0 %; (*open square*) 3.5 %; (*open triangle*) 4.0 %; (*open diamond*) 4.5.0 %; (*open hexagon*) 5.0 %; (*open plus symbol*) 5.5 %. The data was least-squares fitted to Eq. 5.4. Inset shows the variation of exponent m with clay concentration. The age of the sample is 1 year

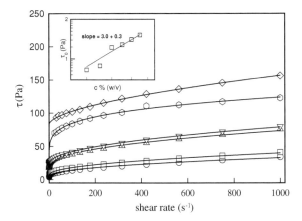

Fig. 5.7 Flow curves for the clay samples of different concentrations 3.0, 3.5, 4.0, 4.5 and 5 % (w/v) from *bottom* to *top* respectively. The solid lines show fitting to the Herschel-Bulkley model. The value of exponent n is 0.45 ± 0.02 and is same for all the curves. Inset shows the variation of yield stress as a function of clay concentration. The power-law exponent determined was $\alpha = 3.0 \pm 0.3$. The age of samples is 1 year

model selected to do the extrapolation. The yield stress τ_0 obtained from the flow curves showed a power-law dependence on clay concentration which is illustrated in Fig. 5.7 and this data could be related to concentration given by Eq. 5.6.

$$\tau_0 \sim c^\alpha, c > c_g; \quad \alpha = 3.0 \pm 0.3 \tag{5.6}$$

The observation of the power-law exponent 3.0 for the gel samples is consistent with the value reported by [19] and [20] for Laponite dispersions. This observation shows that the yield stress dependence on concentration of clay is universal and invariant of aspect ratio of the clay particle.

5.3.4 Cole–Cole Plots

The phase homogeneity in polymer solutions and melts is often deduced from Cole–Cole plot where the imaginary part of the complex viscosity η'' is plotted as function of the real part η'. Typically, in a melt, at very low frequency, viscous behaviour is observed whereas at higher frequencies elastic properties dominate. In this formalism, $\eta^*(\omega) = \eta'(\omega) + i\eta''(\omega)$ and the low and high frequency viscosity values are given by η_0 and η_∞ respectively. The Cole–Cole empirical expression is written as [7]

$$\eta * - \eta_\infty = \frac{(\eta_0 - \eta_\infty)}{\left[1 + (j\omega\tau_{cc})^{1-\alpha}\right]}; \quad 0 < \alpha < 1 \tag{5.7}$$

The aforesaid expression is interpreted as arising from a superposition of several Debye relaxations [7, 30]. The mean relaxation time is given by τ_{cc}. This representation has been used extensively to map homogeneity of soft matter systems and their composites. For a homogeneous phase, the Cole–Cole plot is a perfect semicircle ($\alpha = 0$) with a well defined relaxation time. Any deviation from this shape indicates non-homogeneous dispersion and phase segregation due to immiscibility. Such phases are associated with relaxation time distributions mentioned earlier. Previous studies of materials in which internal processes are accompanied by the loss of energy (dielectric polarization, mechanical deformation etc.) showed that plotting the two components of dynamic viscoelastic characteristics (viscous and elastic modulii) against each other yields an arc shaped curve if the process can be described with a single relaxation time [5, 32]. If the material possesses a relaxation time spectrum, the arc transforms to a semicircle or a skewed semicircle [5, 11, 32]. If more than one process with different relaxation times occur simultaneously, the so called Cole–Cole plot is further modified; i.e. a new semicircle or tail appears [1]. The Cole–Cole plots of MMT dispersions are shown in Fig. 5.8 for our aging samples.

Initially, the dispersions were homogeneous which is evident from the nearly semicircular profile of cole–cole plot. As the sample aged the tail started appearing which indicated network formation in the system [1]. In a different example, the silicate network was detected by the increase in the storage modulus and complex viscosity with decreasing frequency, in the small frequency range of the mechanical spectrum [1, 5, 11, 32]. The formation of a network structure leads to the increase in the elastic component of the modulus and viscosity, and the network obviously deforms with different relaxation times than the homogeneous melt, thus

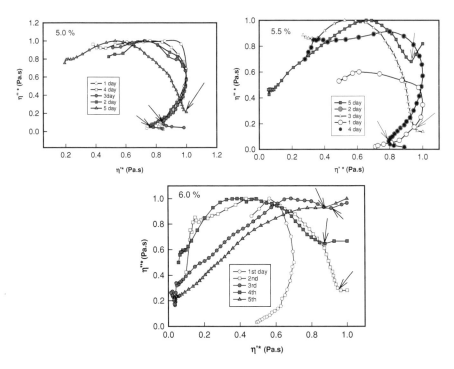

Fig. 5.8 Cole–Cole plots for the specified concentrations of MMT with aging. The data were normalized by dividing η' with η'_{max} and η'' with η''_{max}. The *arrows* indicate the point of upturn

we expect a deviation from a semicircle in the above mentioned representation. The deviation from a skewed semicircle is clearly visible as the sample aged. The change in the shape of the plots indicated the appearance of a new relaxation process, probably the formation of the clay network.

We have employed the rheology to monitor network rigidity of the arrested phase. It was observed that the elastic modulus is higher than the viscous modulus for $c > c_g$ as shown in Fig. 5.9. The samples drawn from the lower part of 2.0 and 2.5 % (w/v) showed viscoelastic behavior whereas the same taken from 1.0 and 1.5 % (w/v) samples showed viscous property.

5.3.5 Light Scattering Experiments

Dynamic light scattering experiments have been performed on dilute samples with concentration $c < c_g$ since the scattering from the concentrated samples will be dominated by multiple scattering. The intensity of scattered light from clay dispersions normalized to initial time ($t = 0$ s) and plotted against waiting time, which is shown in Fig. 5.10. Intensity of the light scattered off the samples showed a plateau

5.3 Results and Discussion

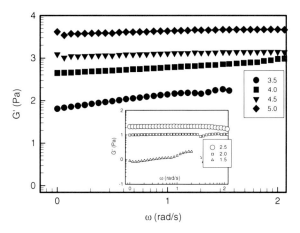

Fig. 5.9 Frequency dependent storage modulii of the MMT dispersions having different concentrations after 2 years of sample preparation. Inset shows the same for samples with $c < c_g$, which are performed on the samples drawn from the lower part of the cell

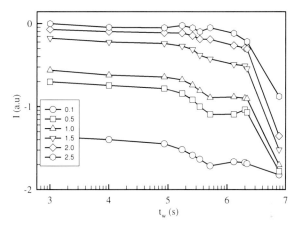

Fig. 5.10 Variation of normalized scattered intensity (normalized at t = 0 s) from clay dispersions below c_g as function of aging time. Clay concentrations are in % (w/v)

in the initial times followed by a gradual decrease and then another narrow plateau followed by a sharp drop. The first plateau indicates that the particle dispersions are stable for a particular time and the interactions did not affect the morphology of the self-assemblies present in the dispersions. A drop in the scattering intensity at later stage is due to the formation of chain-like strands or band like structures. These agglomerations may be reversible due to the presence of second plateau in the later stage. In the last stage there a sharp drop in the scattering intensity due to settling of the clay particles. The drop in the in scattering intensity due to formation of band-like or chain-like structures due to collision of the particles may be a reversible process, while the drop in scattering due to settling is irreversible process (loss of particles from scattering volume).

As the sample aged the size of the aggregates started growing due to EF or FF interactions at lower concentrations. The particles or the aggregates explore the entire phase space and interact and form the aggregates of larger size that

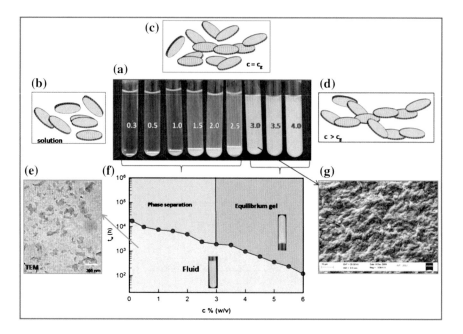

Fig. 5.11 *Phase Diagram of aging MMT dispersions.* **a** Series of MMT dispersions indicating the phase separation states below c_g (= 2.5 %) and the equilibrium gels above c_g. This photograph was taken after 3.5 years. Note that the height of the equilibrium line increases with the clay concentration below c_g. Clay concentrations are in % (w/v). **b–d** Cartoon of self-assembly of clay microstructures. **e** TEM micrograph of MMT in dilute region. **f** Phase diagram of MMT with concentration of clay versus the aging time or the phase separation time. Three regions were clearly identified: stable fluid, phase separation and equilibrium gel. **g** SEM image of MMT dispersion (3 % (w/v))

eventually phase separate due to gravity. The picture that we have observed is quite similar to Laponite system reported recently by [23]. Samples undergo an extremely slow, but clear, phase-separation process into clay-rich and clay-poor phases that are the colloidal analogue of vapor-liquid phase separation. Figure 5.11a depicts the picture of phase separation at lower concentrations below c_g and the formation of stable gels above c_g. Stunningly, the phase separation terminates at a finite but very low clay concentration, above which the samples remain in a homogeneous arrested state, which is called an empty liquid [15, 23].

5.3.6 Observation of Phase Separation and Equilibrium Gels and Phase Diagram

Now the question arises: what is the kind of interactions present that gives rise to network-like self-assembled structures under the present experimental conditions? As reported by [10] for bentonite slurries there are three linear regions in yield stress versus zeta potential data. Two negative slope linear regions were observed where the steepest slope occurred in low pH and a moderate slope was seen in the intermediate pH region. A positive slope was observed for pH > 7. The positive slope region was attributed to heterogeneous charge attraction between the negative face and positive edge. The nature of the platelet edge charge must still be positive in high pH dispersions. The most likely self-assembled microstructure is therefore the "overlapping coins" or the band-like structure (Fig. 5.11b–d). Recently [8] found the aggregates at pH≈9. The fact that aggregates nevertheless form, presumably, is due to the distribution of charges in and around the clay platelets, leading to quadrupole (and higher order multi-pole) interactions between particles. They have found existence of the strand-like structures in MMT suspensions.

In another study, Jonsson et al. [13] modeled the free energy of interaction between two disks each with a 100 positive edge charge sites and 556 negative face charge sites, in edge–face, overlapping coin and staked platelet configurations using a Monte Carlo simulation. This modeling was used as a representation of nanometric clay platelet interactions. They observed that "overlapping coin" configuration produced a global free energy minimum at intermediate salt. In a compression study of sodium bentonite (montmorillonite) slurry, the clay particles initially formed a disordered gel but after the first compression test began forming an "overlapping coin" or parallel plate array structure irrespective of the NaCl concentration ranging from 0.0001 to 0.1 M (See Fig. 5.11b–d). Presence of individual platelets is seen at low solid concentration as observed by TEM micrographs shown in Fig. 5.11e. SEM images show the network like structures in MMT dispersions shown in Fig. 5.11g. Thus we have proposed a phase diagram of extensively aged samples shown in Fig. 5.11f.

The irreversible dehydration transition is similar to the one we have reported in our previous studies [21]. Hardening of gel took place in MMT dispersions in the gel state is reported earlier. The transition temperature did not change with concentration of clay and remained fixed at 45 °C.

5.3.7 Gelation Kinetics in Percolation Formalism

The universal exponents, k and t were obtained from the analysis of relative viscosity and low frequency modulus data using Eqs. 5.1 and 5.2 which prompted us to apply percolation model to explain the kinetics of self-assembly. Just above the gelation transition, it is predicted that the exponent t assumes values $t = 1.7$ for a percolating network, and 2 for a conducting network [27]. Close to the gelation transition, the in phase storage modulus $G'(\omega)$ and out of phase dissipation modulus $G''(\omega)$ exhibit a dispersion relation given by Durand et al. [9]

$$G'(\omega) \sim G''(\omega) \sim \omega^{\delta}; (c \approx c_g) \qquad (5.8)$$

Table 5.1 Comparison of scaling exponents obtained from measurements performed on MMT dispersions

Power-law function	Critical exponent	Classical model	Percolation model	This study (with time in days)		
				1	2	5
$\eta_r \sim \left(\frac{c}{c_g} - 1\right)^k$	k	—	0.7: conducting 1.3: Rouse model 0.75: Zimm model	1.41	1.82	2.06
$G'_0 \sim \left(\frac{c}{c_g} - 1\right)^t$	t	3	1.7	1.55	2.13	3.83
$G'(\omega) \sim G''(\omega) \sim \omega^\delta$	δ	—	0.7	0.53	0.54	0.65

The classical and percolation model exponent values were obtained from [27]

with k, t and δ related through hyper scaling expression [3, 27]

$$\delta = \frac{t}{k+t} \quad (5.9)$$

Theoretical models have shown that in a 3-D system, $k = 0.7$ and $t = 2$ for a conducting network, and $k = 1.3$ and $t = 3$ for a percolating network with Rouse dynamics. Interestingly, the pair, $k = 0.7$ and $t = 2$, and $k = 1.3$ and $t = 3$ yield same value for $\delta \approx 0.7$. Both pair of exponents has been observed for gelling systems [3, 27].

The value for δ obtained from our data is tabulated in Table 5.1. There is a considerable difference in the values of the exponents obtained from experiments and theory, but all the exponents increase with aging of the sample. For instance, value of δ increased from 0.53 to 0.65 with time, which infers that aging was responsible for formation of the self-assembled network. Thus, it can be presumed that the system exhibited anomalous percolation behavior.

5.4 Conclusions

We have undertaken a comprehensive study to map the phase diagram of MMT dispersions extended over a time span of 3.5 years. The system exhibited three phase states: (i) initially stable homogeneous solution (ii) with aging the particles self-assembled and phase separated for concentrations $c < c_g$ and (iii) in the third phase the dispersions entered into an equilibrium gel phase for $c < c_g$, where the self-assembled 3-D networks formed through overlapping coin packing interactions. These samples exhibited aging behavior which was captured through the rheology measurements. Most recently similar observations were made for aging Laponite dispersions, which also exhibited empty liquids, gels and glasses during the course of waiting. It may be concluded that the phenomenon of phase separation and formation of equilibrium gels is ubiquitous in anisotropic colloidal particles associated with inhomogeneous charge distribution. Remarkably the aforesaid phenomenon was independent of platelet aspect ratio. Most recently Meneses-Juarez and co-workers have observed the formation of a liquid phase with vanishing density, called an empty liquid. It was clearly inferred that such arrested empty liquid states are formed under reduced valence and low coordination number conditions in aging dispersions spontaneously due to the patchy distribution of charge on the platelet surface. Thus, the comprehensive understanding of the kinetics of dynamic arrest and phase separation in the discotic colloidal suspensions continues to remain a challenging and poorly explored problem.

References

1. A. Abranyi, L. Szazdi, B. Pukanszky Jr, G.J. Vancso, B. Pukanszky, Formation and detection of clay network structure in poly(propylene)/layered silicate nanocomposites. Macromol. Rapid Commun. **27**, 132–135 (2006)
2. M. Adams, Z. Dogic, S.L. Keller, S. Fraden, Entropically driven microphase transitions in mixtures of colloidal rods and spheres. Nature **393**, 349–352 (1998)
3. A. Aharony, D. Stauffer, *Introduction to percolation theory* (Taylor and Francis, London, 1994)
4. U. Brandenburg, G. Lagaly, Rheological properties of sodium montmorillonite dispersions. Appl. Clay Sci. **3**, 263–279 (1988)
5. P.J. Carreau, M. Bousmina, A. Ajji, Rheological properties of blends: facts and challenges. Prog. Pacific Polym. Sci **3**, 25–29 (1994)
6. B.H. Cipriano, T. Kashiwagi, X. Zhang, S.R. Raghavan, A simple method to improve the clarity and rheological properties of polymer/clay nanocomposites by using fractionated clay particles. ACS Appl. Mater. Interfaces **2009**(1), 130–135 (2009)
7. K.S. Cole, R.H. Cole, Dispersion and absorption in dielectrics I. alternating current characteristics. J. Chem. Phys. **9**, 341–351 (1941)
8. Y. Cui, C.L. Pizzey, J.S. Van Duijneveldt, Modifying the structure and flow behaviour of aqueous montmorillonite suspensions with surfactant. Phil. Trans. R. Soc. A **371**, 20120262 (2013)
9. D. Durand, M. Delsanti, M. Adam, J.M. Luck, Frequency Dependence of Viscoelastic Properties of Branched Polymers near Gelation Threshold. Europhys. Lett. 3:297 (1987)
10. R. Goh, Y.K. Leong, B. Lehane, Bentonite slurries—zeta potential, yield stress, absorbed additive and time-dependent behavior. Rheol. Acta **50**, 29–38 (2011)
11. S. Havriliak, S. Negami, A complex plane analysis of α-dispersion in some polymer systems. J. Polym. Sci. C **14**, 99–117 (1966)
12. W. Herschel, R. Bulkley, Konsistenzmessungen von gummi-benzollösungen. Colloid Polym. Sci. **39**, 291–300 (1926)
13. B. Jonsson, C. Labbez, B. Cabane, Interaction of Nanometric Clay Platelets. Langmuir **24**, 11406–11413 (2008)
14. G. Lagaly, Principles of flow of kaolin and bentonite dispersions. Appl. Clay Sci. **4**, 105–123 (1989)
15. E. Meneses-Juarez, S. Varga, P. Orea, G. Odriozola, Towards understanding the empty liquid of colloidal platelets: vapour-liquid phase coexistence of square-well oblate ellipsoids. Soft Matter **9**, 5277–5284 (2013)
16. L.J. Michot, I. Bihannic, K. Porsch, S. Maddi, C. Baravian, J. Mougel, P. Levitz, Phase diagrams of wyoming na-montmorillonite clay influence of particle anisotropy. Langmuir **20**, 10829–10837 (2004)

17. E. Paineau, L.J. Michot, I. Bihannic, C. Baravian, Aqueous Suspensions of Natural Swelling Clay Minerals. 2. Rheological Characterization. Langmuir 27, 7806–7819
18. T. Permien, G. Lagaly, The rheological and colloidal properties of bentonite dispersions in the presence of organic compounds. 1: flow behavior of sodium-bentonite in water—alcohol. Clay Miner. **29**, 751 (1994)
19. F. Pignon, J.M. Piau, A. Magnin, Structure and pertinent length scale of a discotic clay gel. Phys. Rev. Lett. **76**, 4857–4860 (1996)
20. R.K. Pujala, H.B. Bohidar, Slow dynamics, hydration and heterogeneity in laponite dispersions. Soft Matter **9**, 2003–2010 (2013)
21. R.K. Pujala, N. Pawar, H.B. Bohidar, Universal sol state behavior and gelation kinetics in mixed clay dispersions. Langmuir **27**, 5193–5203 (2011)
22. M.M. Ramos-Tejada, F.J. Arroyo, R. Perea, J.D.G. Duran, Scaling behavior of the rheological properties of montmorillonite suspensions: correlation between interparticle interaction and degree of flocculation. J. Colloid Interface Sci. **235**, 251–259 (2001)
23. B. Ruzicka, E. Zaccarelli, L. Zulian, R. Angelini, M. Sztucki, A. Moussaid, T. Narayanan, F. Sciortino, Observation of empty liquids and equilibrium gels in a colloidal clay. Nature Mater **10**, 56–60 (2011)
24. M. Segad, B. Jonsson, B. Cabane, Tactoid formation in montmorillonite. J. Phys. Chem. C **116**, 25425–25433 (2012)
25. A. Shalkevich, A. Stradner, S.K. Bhat, F. Muller, P. Schurtenberger, Cluster, glass and gel formation and viscoelastic phase separation in aqueous clay suspensions. Langmuir **23**, 3570–3580 (2007)
26. P. Shankar, J. Teo, Y. Leong, A. Fourie, M. Fahey, Adsorbed phosphate additives for interrogating the nature of interparticle forces in kaolin clay slurries via rheological yield stress. Adv. Powder Technol. **21**, 380–385 (2010)
27. D. Stauffer, A. Coniglio, A. Adams, Polymer networks. Adv. Polym. Sci. **44**, 103–158 (1982)
28. F.M. Van der Kooij, K. Kassapidou, H.N.W. Lekkerkerker, Liquid crystal phase transitions in suspensions of polydisperse plate-like particles. Nature **406**, 868–871 (2000)
29. G.J. Vroege, D.M.E. Thies-Weesie, A.V. Petukhov, B.J. Lemaire, P. Davidson, Smectic liquid-crystalline order in suspensions of highly polydisperse goethite nanorods. Adv. Mat **18**, 2565–2568 (2006)
30. T.C. Warren, J.L. Schrag, J.D. Ferry, Infinite-dilution viscoelastic properties of poly-γ-benzyl-L-glutamate in helicogenic solvents. Biopolymers **12**, 1905–9015 (1973)
31. T. Yalcin, A. Alemdar, O.I. Ece, N. Gungor, The viscosity and zeta potential of bentonite dispersions in presence of anionic surfactants. Mater. Lett. **57**, 420–424 (2002)
32. Q. Zheng, M. Du, B. Yang, G. Wu, Relationship between dynamic rheological behavior and phase separation of poly (methyl methacrylate)/poly (styrene-co-acrylonitrile) blends. Polymer **42**, 5743–5747 (2001)

Chapter 6
Sol State Behavior and Gelation Kinetics in Mixed Nanoclay Dispersions

Abstract This chapter investigates sol and gel state behavior, in aqueous salt free dispersions, of clays Laponite (L) and Na Montmorillonite (MMT) at various mixing ratios (L:MMT = r = 1:0.5, 1:1 and 1:2). Percolation model have been used to understand the gelation kinetics. This is the first systematic study in the literature on the mixed clay dispersions.

6.1 Introduction

In recent years, the suspensions of the colloidal plate-like clay particles have been the subject of intense investigations. Aqueous clay suspensions are fascinating model systems to study fundamental problems in colloid science and have recently attracted particular attention from scientists belonging to various disciplines. An interesting feature of clay suspension is the existence of interactions at different length scales combined with the anisotropic structure of the particles, which lead to a variety of structural and dynamical phenomena such as orientation, ordering and structuring, and clustering (like house-of-cards) [35] as a function of the particle concentration and of the ionic strength. Recent advances in the description and understanding of aggregation, and glass formation have led to a wealth of new information in the general area of dynamical arrest and viscoelastic phase separations [32]. The gelation of dispersions of clay particles is both industrially important, as well as scientifically challenging [20, 25]. Gels have applications in drilling fluids, paints, ceramic additives, and in cosmetic and pharmaceutical formulations. Clays having different aspect ratio are routinely used as the additives in making customized polymer nanocomposites [19].

Over more than 50 years, various mechanisms of gelation and suspension of aggregated structures have been proposed for clays. Van Olphen proposed that clay dispersions may form three dimensional aggregated structures; the so-called [35] "house of cards". This structure was believed to arise from electrostatic attraction between oppositely charged double layers at the edges and faces of the particles as a result of the different chemical composition of the face and edge surfaces. The faces have a constant negative charge due to substitution within the lattice. The edges are like oxide surfaces; therefore, they may be positively or negatively charged depending on the pH of the suspension. All these years the research on clays was largely focused on studying the properties like the morphology [4], anisotropy [16], gelation [11], long-term gelation [31], fractal growth [26] and arrested disordered state of colloidal glass [30, 31] mainly of the single component systems. Many phase diagrams [21, 31, 29] have been proposed based on the visual observation [20], viscoelastic properties [27] and optical polarization [14] experiments. Little attention has been paid to the study of binary clay systems and their phase stability behavior.

The system of clay with different aspect ratio behaves differently showing changes in the phase states, optical, viscoelastic and rheological properties. Our motivation was two fold: how the system behaves and gels when the colloidal discs of different aspect ratio mix in various proportions. To the best of our knowledge, this is the first systematic and comprehensive study made on the binary mixture of the clays. For this study, we used two clays of different aspect ratio. The two clays are sodium montmorillonite (MMT) having large aspect ratio ($\approx$250) and Laponite with small aspect ratio ($\approx$30). The phase diagrams of Laponite and MMT were studied in Chaps. 3, 5 respectively. Individual dispersion behavior of both the clays used in this study is well known. This raised the pertinent question: How would the mixing of the two clays affect the gelation features of the resultant cogels, and their thermo-mechanical properties? We address these issues herein. Considering the fact that clays are routinely used as mechanical property enhancers, study of the temperature dependent behaviour of clay preparations is of significant importance. We used the term cogel to describe this arrested phase made of Laponite-MMT in this chapter.

6.2 Sample Preparation

Characterization of Laponite and MMT was described in Chap. 2. Gelling concentration of Laponite and MMT is known to be 2 and 3 % (w/v) respectively (discussed later). The fractionation of MMT was done as described in Chap. 5. The desired samples of individual dispersions were prepared by dispersing the clay powder in deionized water at three different concentrations 1.5, 2.0 and 2.5 % (w/v) followed by vigorous stirring for 2 h and then the aliquots were filtered through 0.45 μm Millipore filter. Different concentration (4, 3, 2 %) w/v of Laponite and MMT dispersion were prepared by dissolving Laponite and MMT separately in deionized water at room temperature and mixing them vigorously for 2 h using magnetic stirrer. Solutions of clay for different mixing ratio (1:0.5, 1:1 and 1:2) were prepared by mixing 4, 3 and 2 % L (w/v) with 2, 3 and 4 % MMT (w/v) respectively in equal volumes to generate optically clear binary mixtures. All the

6.2 Sample Preparation

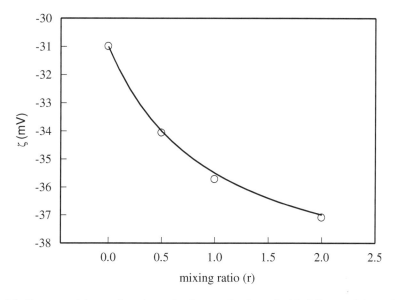

Fig. 6.1 Zeta potential or surface charge for the complex formed with different mixing ratio of two clays. *Solid line* is fitting of data to Eq. 6.1. *Symbol size* represents experimental error

solutions were filtered through 0.45 μm Millipore filter paper to remove large size fractions [28]. Initially the samples showed pH ~ 9.0. After an aging time (~7 h) samples entered the gel state. Enough time (2 days) was given to the samples to stabilize in the gel phase. Studies were performed on these gel samples using rheology and XRD techniques. All the experiments were performed at room temperature $T = 25\ °C$.

6.3 Result and Discussion

6.3.1 Sol State Behavior

The zeta potential data pertaining to the individual clay components and their mixtures was measured which is shown in Fig. 6.1. The complex with higher concentration of Laponite was seen to carry more negative surface charge. As mentioned earlier in Chap. 2, the reported values for zeta potential for Laponite and MMT are −40 and −30 mV respectively. While the measured zeta potential value for MMT is close to the literature value the same for Laponite differs by ≈30 % which is not unusual. Diluted clay dispersions were used for the zeta potential measurements.

As the mixing ratio increased from 1:0.5 to 1:2 the zeta potential became less negative. Interestingly, the zeta potential (Fig. 6.1) of the mixed colloids (ζ_{L-MMT})

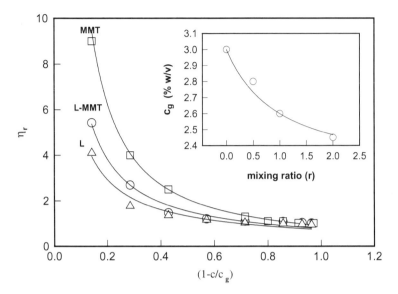

Fig. 6.2 Plot of relative viscosity of clay suspensions as function of concentration. The data was fitted to Eq. 6.2 (solid lines) which yielded $k = 0.8$, 1.0 and 1.2 for Laponite, Laponite-MMT ($r = 1:1$) and MMT dispersions respectively. The *inset* shows the variation of gelation concentration with mixing ratio and the solid line is fitting of data to Eq. 6.2 ($\chi^2 = 0.97$). Symbol size represents experimental error

was found to be a linear combination of their constituent colloids (ζ_L and ζ_{MMT}) given by

$$\zeta_{L-MMT} = (r\zeta_L + \zeta_{MMT})/(1 + r) \tag{6.1}$$

The observed effect can be attributed to charge neutralization which will happen when there is edge-face binding between Laponite and MMT particles. This is indicative of preferential binding between face of MMT with the edge of Laponite in the aggregates formed. The typical number density was $\approx 10^{18}$ particles/cc and 10^{16} particles/cc for Laponite and MMT respectively.

The zero shear viscosity of the clay dispersions, as function of concentration, was measured using a digital Vibro viscometer. The data is shown in Fig. 6.2.

The relative viscosity data could be described by the power-law function

$$\eta_r \sim \left(1 - \frac{c}{c_g}\right)^{-k} ; \quad (c < c_g) \tag{6.2}$$

where $\eta_r = \eta/\eta_0$, η is the viscosity of dispersion and η_0 is solvent viscosity. The clay concentration is c and its gelation concentration is c_g. The sol–gel transition is characterized by an exponent, $k = 0.9$ for colloidal gels [31, 34]. The least squares fitting yielded $c_g = 2.2$ and 3.0 % (w/v) for Laponite and MMT samples. For the mixed sols, it was 2.4, 2.6 and 2.9 % (w/v) for $r = 1:0.5$, 1:1 and 1:2 samples respectively. The exponent, $k = 0.8$, 1.2 and 1.1 ± 0.1 for Laponite, MMT,

6.3 Result and Discussion

Table 6.1 Comparison of physico-thermal properties of clay sols, their gels and cogels determined at room temperature (25 °C)

Parameter	Laponite system	Montmorillonite system	Cogel system
C_g (% w/v)	2.2	3.0	2.4; r = 1:0.5 2.6; r = 1:1 2.9; r = 1:2
Zeta potential (mV)	−40	−30	−37; r = 1:0.5 −36; r = 1:1 −34; r = 1:2
R_{app} (nm)	18	120	–
Basal spacing (nm)	1.47	1.27	1.28; r = 1:0.5 1.26; r = 1:1 1.26; r = 1:2
G_0' (at 1 rad/s) (Pa)	680	670	5,000; r = 1:0.5 10,500; r = 1:1 2,000; r = 1:2
T_c (°C)	45	50	56; r = 1:0.5 63; r = 1:1 56; r = 1:2
Exponent, γ ($G_0' \sim (T_c-T)^\gamma$)	0.25	0.55	0.40 for all r

Laponite-MMT (r = 1:0.5, 1:1 and 1:2) dispersions respectively. The invariance of the exponent k with mixing ratio r indicates universality of sol–gel transition. The relative viscosity, at low concentration is directly related to the intrinsic viscosity, is a function of particle size and shape. Thus, it was concluded that small clusters of Laponite-MMT were present in the dispersion that were relatively more symmetric as compared to MMT colloids. Such a description is supported by the zeta potential data shown in Fig. 6.1. Such clusters thrive on short length scales and owe their origin to edge-face electrostatic interaction between pairs of colloidal particles. As the concentration approached c_g large interconnected colloidal networks are formed that make the sol viscosity diverge.

It was interesting to observe that the gelation concentration of the cogel ($c_{gL\text{-}MMT}$) was a linear combination of the gelation concentrations of Laponite (c_{gL}) and MMT (c_{gMMT}) given by

$$c_{gL\text{-}MMT} = (rc_{gL} + c_{gMMT})/(1+r) \tag{6.3}$$

Equations 6.1 and 6.3 were identical indicating the universal nature of the sol state of the Laponite-MMT mixed dispersions. The sol state properties were summarized in Table 6.1.

6.3.2 Gel State Properties of Mixed Clay Dispersions

The samples remained in the sol form initially, but with the passage of time the interaction between Laponite and MMT ensues and both short and long range orders develop. The net charge on the Laponite-MMT clusters was negative which

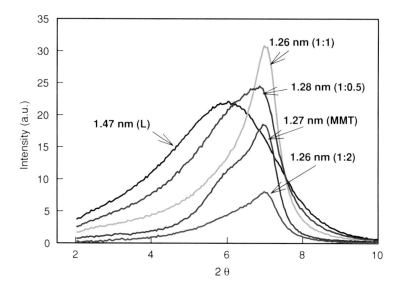

Fig. 6.3 XRD profile for individual clays and the complex formed with different mixing ratios of the two clays, Laponite:MMT = 1:1, 1:0.5, 1:2. The numbers designate inter planar spacing

facilitates attractive short-range but repulsive long-range interactions [34]. Thus, the formation of gel-like structures purely through repulsive interactions (Wigner glass phase) becomes a reality. This conclusion was derived based on the phase diagram proposed by [34] and the data presented in Fig. 6.2. Further studies of the samples were done on the gel phase.

Complete XRD profile before and after mixing of the two clays in different mixing ratio was recorded and the same was presented in Fig. 6.3. It clearly showed the small angle XRD pattern in the range $2\theta = 2-10^0$. These figures showed the absence of Laponite signature in the cogel diffraction profiles. Using these XRD data, inter planar spacing was determined for cogels. In case of MMT, this spacing was close to 1.27 nm while the Laponite inter planar spacing was 1.47 nm. Interconnected networks prevailing over different length scales are formed in the cogel when the binding ratio is altered, regardless, one observed XRD peaks dominated by MMT samples as shown in Fig. 6.3. The sharpness of the peaks is determined by the homogeneity and regularity of the microscopic arrangement of the network structure. The 1:1 sample exhibited the presence of very homogeneous network, which will be discussed later, implying narrow peak with enhanced amplitude. Cogels made with other mixing ratios had significant amount heterogeneity in their network structures and such embodiment gave rise to incoherent diffraction, i.e. broad peaks with low amplitude. This has been the case in Fig. 6.3 data.

No change in inter planar spacing of MMT after mixing with Laponite in the co- gels revealed that Laponite particles did not get intercalated in between the MMT planes. It has been reported [8] that the basal spacing in MMT is 1.4 nm

6.3 Result and Discussion

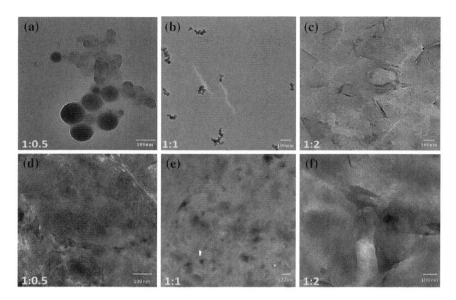

Fig. 6.4 **a, b** and **c** show TEM images of Laponite and MMT mixed in 1:2, 1:1, 1:0.5 ratios below their gelation concentrations and **d, e** and **f** show TEM images of corresponding cogels. Scale bar 100 nm

which corresponds to the XRD peak at $2\theta = 6.4$. Present data locates this at $2\theta = 7.0$ and attributes ≈ 1.3 nm to the basal spacing which is not too different from what is reported in the literature. In the cogels, the XRD peak position was still observed at the same 2θ value and the basal spacing did not change significantly. However, the half-width and peak heights changed significantly clearly implying finite interaction between Laponite and MMT colloids.

Figure 6.4 shows the TEM images of the cogel samples for three mixing ratios ($r = 1:0.5$, 1:1 and 1:2). Figure 6.4a–b depict the sol state pictures of these samples where polydisperse clusters of Laponite in the MMT background are visible. In Fig. 6.4c this is not discernible. The corresponding cogel pictures are shown in Fig. 6.4d–f, it shows homogeneous dispersions in the gel pertaining to $r = 1:1$ sample, the other two samples are observed to be associated with considerable heterogeneity. These figures clearly reveal that cogel made with excess MMT or Laponite yield non-homogeneous samples.

6.3.3 Visco-Elastic Properties

We performed light scattering experiments on cogel samples after 48 h of preparation and found that the baseline of the correlation functions was quite stable, implying fast aging process had already saturated. The slower aging process

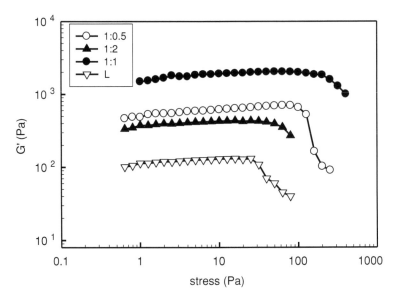

Fig. 6.5 Variation of elastic modulus as a function of stress applied to the complex with different mixing ratios to know the viscoelastic linear domain in the gel samples. Solid lines are guide to eye. Symbol size represents experimental error

occurs over a time period of weeks which was examined by Baghdadi et al. [5] in detail. All our results were obtained within 72 h of sample preparation. Thus, these were not affected by slow aging dynamics. Effect of mixing ratio on the viscoelastic behaviour of the binary gels can be used to map the morphological microstructure of the networks. This was achieved by studying the effect of stress and temperature on the gels formed with different mixing ratio of the two clays.

(a) **Effect on yield stress**

The shear stress amplitude sweep test performed on the binary gels is shown in Fig. 6.5. It is observed that 1:1 complex gel has more extended linear viscoelastic region where the storage modulus, G', values remain relatively constant over a wide range of applied shear stress. The deviation from linear viscoelastic region in 1:1 gel started at higher stress compared to other gels. The stress at which the deviation started gave the indication of yield stress. Hence, the yield stress or network strength was found to be maximum for the $r = 1:1$ cogel. The G' value for this gel was an order of magnitude larger than the same of Laponite gel.

Figure 6.6 showed the transition temperatures and yield stress (derived from Figs. 6.5 and 6.7) for cogels formed with different mixing ratios. These figures clearly indicated that gel with equal amount of both clays (1:1) had best properties like highest yield stress and highest transition temperature.

Pignon et al. [27] measured yield stress in Laponite dispersions and concluded that a characteristic length scale of the structure of thixotropic colloidal clay suspension

6.3 Result and Discussion

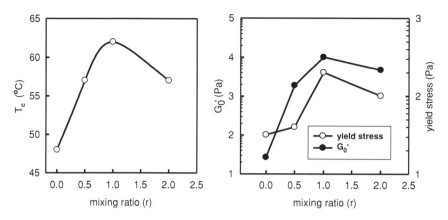

Fig. 6.6 Variation of transition temperature (*left panel*) and yield stress (*right panel*) for the complex formed with different mixing ratios. *L* and *M* represent Laponite and MMT respectively. Solid lines are guide to eye. Symbol size represents experimental error

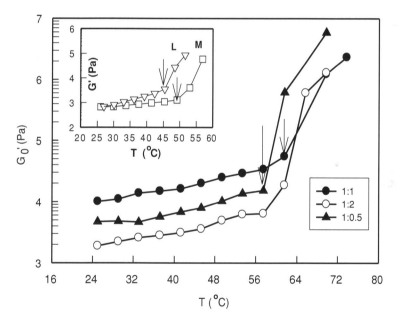

Fig. 6.7 Temperature dependent variation of low frequency storage modulus of various gel samples. Inset shows the same for individual clays: Laponite (L) and MMT (M). *Arrows* point to transition temperatures. *Solid lines* are guide to eye. *Symbol size* represents experimental error

could be established through static structure factor studies. The length scale was dependent on the volume fraction of the clay and the yield stress could be linked to the fractal dimension of aggregates formed in the system which increased from 1 to 1.8 as the volume fraction reached a critical value. This observation suggested that the gels comprised of large micro sized aggregates that contributed to rheological profile of a given discotic gel. In a related study, [7] reported that fractal-like organization of clay particles in Laponite gels was not the case and the there was no correlation between fractal formation and viscoelastic properties of the gel. The Laponite network was conceived as a colloidal glass, rather than a conventional gel. They further argued that the fractal-like aggregates observed in earlier studies could possibly owe their origin to incomplete dissolution of Laponite in water. The rheology data implied that the arrested phase comprised a solvated polymer network-like phase with a large storage modulus (Fig. 6.6). Such a large storage modulus is unlike of a colloidal dispersion devoid of connectivity, which indicates the existence of highly crosslinked networks inside the medium. The colloidal cogel material was associated with much higher tensile strength when made with mixing ratio 1:1. The storage modulus is a measure of elastic energy stored in the material. Thus, a perfectly structured network will be associated with a high G' value. This gel-like behaviour was further probed through thermal characterization.

(b) **Thermal characterization**

There is considerable debate in the literature pertaining to the question: what is the physical designation of the arrested phase of clay dispersions? This question is best resolved through thermal characterization of the samples. The influence of temperature on the same cogels was studied by a temperature sweep experiment where the storage modulus was probed under temperature scan (Fig. 6.7). The cogels showed anomalous thermal behavior with T_c increasing by $\approx$13 °C for 1:1 cogel as compared to that of MMT. Heating the gels of individual clay samples and their cogels resulted in an increase in the elastic modulus and an abrupt change in increase was observed at a characteristic temperature. The increase in elastic modulus with increasing temperature is attributed to the release of solvent from the network and their associated reorganization. An additional contribution arises from dehydration effect. The solvent water is present in the gel in two different physical states [6], namely: (i) the free water trapped between planes of clay material and (ii) hydrogen bonded coordinated water. The free water gets released in the temperature range 20–100°C resulting in considerable weight loss of the sample and this could lead to reduction in inter planar spacing. The coordinated water loss happens at much higher temperature (>100 °C). Thus, the dehydration at lower temperature, as observed in the present case, would lead to higher rigidity of the gel material as the gel network de-swells in a confined geometry.

The data shown in Fig. 6.7 clearly identify the temperature, T_c where a sharp increase in the value of G' occurred. Thus, unlike polymer gels that reveal a melting transition, the discotic gels were found to be exhibiting a hardening transition, a behavior that was again universal. This allows scaling of the data and depicting it alternatively as given in Fig. 6.8.

The storage modulus data shown in Fig. 6.8 manifests the temperature dependent changes that could be described through a power-law scaling given by

$$G'_0 \sim (T_c - T)^{-\gamma} \tag{6.4}$$

6.3 Result and Discussion

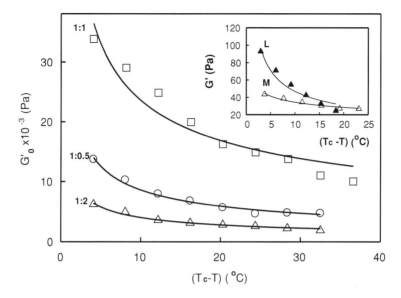

Fig. 6.8 Temperature dependence of low frequency storage modulus (G_0') of various clay gels. Numbers near the *curves* represent mixing ratio. This is a scaled reproduction of data shown in Fig. 6.7. *Solid lines* show the power law fitting: $G' \sim G_0(T_c - T)^{-\gamma}$ ($\chi^2 = 0.98$). Symbol size represents experimental error. See text for details

Interestingly, $\gamma = 0.40 \pm 0.05$ was found to be invariant of composition of the cogel. However, for MMT and Laponite it was 0.25 and 0.55 respectively. The data indicated that when the two components were available in relatively same quantity the network generated possesses excellent rigidity. In the repulsive interaction model, higher surface charge should generate a stronger gel. In the pristine, clay gels storage modulus of the associated networks had G' values two decades smaller than the cogels. This was regardless of the fact that MMT has a surface charge comparable to that of Laponite-MMT system. Thus, the repulsive interaction model could not describe the gel rigidity adequately. It was argued that there must be colloidal network-like structures in the gel phase that contribute to the high storage modulus values observed.

The storage modulus is a measure of elastic free energy stored per unit volume of a characteristic viscoelastic network of size, ξ. The sample heterogeneity can be deduced from the viscoelastic length (ξ) persisting inside a given phase. This parameter estimates the typical separation between colloid rich and poor regions inside the sample at temperature, T and relates to low frequency storage modulus, G'_0, as [3]

$$G'_0 \approx k_B T / \xi^3 \qquad (6.5)$$

This data is plotted in Fig. 6.9 where a monotonous decrease in this length was observed as the temperature was increased. In the repulsive gel model (Wigner glass)

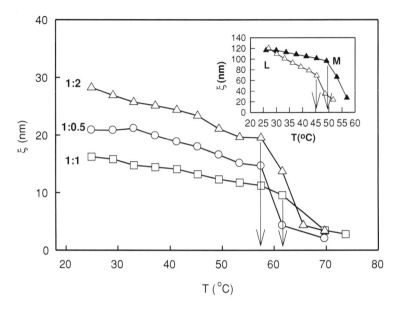

Fig. 6.9 Plot of viscoelastic length, ξ, as function of temperature for various clay gels and cogels. *Arrows* indicate transition temperature, T_c. *Solid lines* are guide to eye. *Symbol size* represents experimental error

of [34] such a situation would increase repulsive interactions; thereby, increasing the gel rigidity. All the physico-thermal properties of the mixed clay dispersions were summarized in Table 6.1.

6.3.4 Gelation Kinetics in Percolation Formalism

The ionic strength and clay concentration define the arrested phase of the dispersion when $c > c_g$ which was described extensively in the literature [17, 23, 24, 28, 34]. For ionic strength >1 mM, there is considerable screening of the electrostatic potential that leads to strong Coulombic attraction between the oppositely charged faces of clay particles and gelation becomes a possibility. On the other hand, electrostatic potential of the clay particles cause strong interparticle repulsion in salt-free and low salt dispersions. Thus, the dispersion can lead to glass formation, which is normally called a Wigner glass in the literature [34]. Such a situation should prevail in our case. Let us examine the data on hand. Since, percolation theory is a well developed concept that successfully accounts for gelation in many systems [2], it was considered worthwhile to apply this model to the present system. Since, $r = 1:1$ cogel was found to be a homogeneous preparation, this system was probed extensively.

The gelation transition used to be described through the classical theory first conceptualized by Flory-Stockmayer [1, 2, 9, 10, 13, 33]. Though this model

6.3 Result and Discussion

accounts for the gelation phenomenon successfully in various systems, it relies heavily on the geometrical properties and inadequately incorporates any dynamics on its own. In this model, the critical growth of connectivity is directly interpreted in terms of percolation transition. The main drawback of this theory is that the model is not based on any periodic lattice structure. Thus, the predicted critical exponents are independent of space dimensionality and monomer functionality. As a result, hyperscaling cannot be applied to these critical exponents [33].

Many of these short comings were addressed in percolation theory which provides a statistical description to the phenomena of gelation. The detailed theory concludes that a percolation transition (gelation) is like a second-order phase transition with c playing the role of temperature (like a liquid–gas phase transition at the critical point) with well characterized universal behavior [1, 2, 10, 13, 33]. For $|c - c_g| \to 0$ one observes universal scaling with characteristic exponents. As per this formalism, as the gelation transition is approached; say by increasing the concentration to gelation concentration, the sol viscosity (η_r), low frequency shear modulus (G_0') and apparent cluster size (R) exhibit characteristic scaling exponents. The viscosity behavior was already described through Eq. 6.2 where the critical exponent was k. The low frequency storage modulus, G_0' scales with concentration as

$$G_0' \sim \left(\frac{c}{c_g} - 1\right)^t; \quad (c > c_g) \tag{6.6}$$

Just above the gelation transition it is predicted that the exponent $t = 1.7$ for a percolating network, and 2 for a conducting network [33]. Close to the gelation transition, the in phase storage modulus, $G'(\omega)$ and out of phase dissipation modulus, $G''(\omega)$, exhibit a dispersion relation given by [13]

$$G'(\omega) \sim G''(\omega) \sim \omega^\delta; \quad (c \approx c_g) \tag{6.7}$$

with k, t and δ related through hyper scaling expression [2, 33]

$$\delta = \frac{t}{k + t} \tag{6.8}$$

Theoretical models have shown that in a 3-D system, $k = 0.7$ and $t = 2$ for a conducting network, and $k = 1.3$ and $t = 3$ for a percolating network with Rouse dynamics. Interestingly, the pair, $k = 0.7$ and $t = 2$, and $k = 1.3$ and $t = 3$ yield same value for $\delta \approx 0.7$. Both pair of exponents has been observed for gelling systems [2, 33]. In addition, for a percolating network below gel point, only finite size clusters are present and as $c \to c_g$ clusters grow following a power-law behavior

$$R \sim \left(1 - \frac{c}{c_g}\right)^{-\upsilon}; \quad (c < c_g) \tag{6.9}$$

R is apparent Stokes radius of the cluster. For a percolating cluster in 3-D, one expects $\upsilon = 0.85$. Note that Eqs. 6.2, 6.6, 6.7 and 6.9 introduce the characteristic

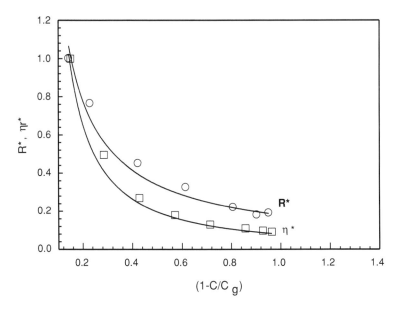

Fig. 6.10 Variation of normalized viscosity (η^*) and apparent cluster size (R^*) as function of concentration. The *curves* could be universally described by the power-law: η^* and $R^* \sim (1-c/c_g)^{-k,\upsilon}$ with $k = 1.2$ and $\upsilon = 0.8$ ($\chi^2 > 0.98$). Symbol size represents experimental error

critical exponents (k, t, δ and υ) with well defined values, and their origin is soundly based on rigorous theoretical calculations and computer simulations data [1, 2, 9, 10, 13, 33].

DLS studies, performed on the gelling sol very close to gelation concentration, revealed that the dynamic structure factor comprised two relaxation modes. The fast mode relaxation mode was diffusive and gave an apparent Stokes radius, R which was a measure of cluster size of the colloidal network evolving in the sol. The slow mode relaxation behaviour was more interesting and will be discussed in Chap. 7.

In our studies, the viscosity exponent, k, of the gelling sol was found to be 1.2 for the cogel (0.8 for Laponite, [31] determined $k = 0.9$; and 1.2 for MMT) as shown in Fig. 6.2. The normalized viscosity $(\eta^* = \eta_r(c)/\eta_r(c_g))$ and apparent cluster size $(R^* = R(c)/R(c_g))$ were plotted as function of concentration in Fig. 6.10. As gelation concentration was approached the sol state properties diverged strongly. The data shown in Fig. 6.10 could be least squares fitted to the general power-law expressions that give the exponent values: $k = 1.1$ and $\upsilon = 0.8$.

Figure 6.11 depicts the plot of low frequency storage modulus, G_0' as function of concentration which yields $t = 1.5$ (Eq. 6.6). Close to the gelation concentration, $G'(\omega)$ and $G''(\omega)$ exhibited interesting behaviour as shown in Fig. 6.12. It was obvious from the data presented in Fig. 6.12 that $G'(\omega)$ and $G''(\omega)$ scale with frequency, ω, with exponents δ' and δ'', and $\delta' \neq \delta''$. The fitting of data to Eq. 6.7

6.3 Result and Discussion

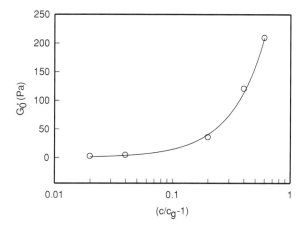

Fig. 6.11 Plot of low frequency storage modulus (G_0') as function of $(c/c_g - 1)$ for $r = 1{:}1$ cogel. The data was fitted to $G_0' \sim (c/c_g - 1)^t$ that produced $t = 1.5$ (Eq. 6.8) ($\chi^2 = 0.98$). Symbol size represents experimental error

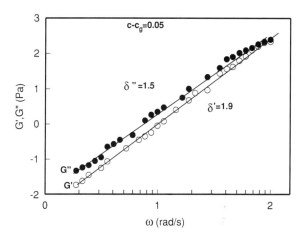

Fig. 6.12 Plot of storage (G') and loss (G'') modulii as function of frequency close to gelation concentration, $c-c_g = 0.05$ for $r = 1{:}1$ cogel. The data was fitted to $G' \sim \omega^{\delta'}$ and $G'' \sim \omega^{\delta''}$ which produced $\delta' = 1.9$ and $\delta'' = 1.5$ (chi-square >0.98). Symbol size represents experimental error

yielded $\delta' = 1.9$ and $\delta'' = 1.5$. This was the only observation where the prediction of percolation theory was at variance with experimental data. The theory necessitates that measurement were to be taken as close to gelation concentration as possible, may be in our studies this condition was not met adequately. However, if we use our experimental values for $k = 1.1$ and $t = 1.5$, we could get $\delta = 0.56$.

Considering, the experimental difficulty associated in the evaluation of the critical exponents, this difference of 20 % between the two values of δ can be ignored. In the absence of any other theory, it was thought appropriate to apply percolation model to the observed results. This exercise yielded well defined exponents though deviations, as expected occurred. Figure 6.13 depicted the change in δ' as function of clay concentration and, interestingly, at $c \approx c_g$, δ' value dropped from 2 to 0.8 indicating the onset of a solid-like behavior (gelation). Thus, below c_g, the dispersion was alike a viscous liquid (fully Maxwellian) while above c_g it behaved as a soft solid.

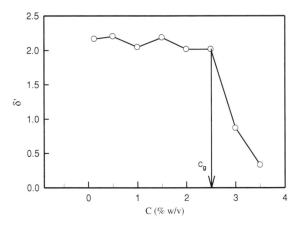

Fig. 6.13 Plot of exponent δ' as function of clay concentration keeping $r = 1:1$ constant. Notice the sharp drop in the value of the exponent as $c \to c_g$. The dispersion transits from a viscous liquid to a solid-like phase. *Solid line* is guide to eye

Table 6.2 Comparison of critical exponents obtained from measurements performed on Laponite: MMT $= 1:1$ gelling sol

Power-law function	Critical exponent	Classical model	Percolation model	This study
$R \sim \left(1 - \frac{c}{c_g}\right)^{-\upsilon}$	υ	0.5	0.88	0.8
$\eta_r \sim \left(1 - \frac{c}{c_g}\right)^{-k}$	k	–	0.7: conducting 1.3: rouse model 0.75: Zimm model	1.1
$G'_0 \sim \left(\frac{c}{c_g} - 1\right)^t$	t	3	1.7	1.5
$G'(\omega) \sim G''(\omega) \sim \omega^\delta$	δ	–	0.7	0.56 (Eq. 6.8)

Note that there is considerable matching between the theoretical predictions of percolation theory and experimental results. The classical and percolation model exponent values were obtained from [33]

Following the early gelation mechanism proposed by Flory-Stockmayer, numerous attempts have been made to seek for a universal gelation model. Colloidal gelation of clay particles is unique in the sense that these exhibit strong ionic strength dependent pair-wise interactions. In addition, the mutual orientation of the clay particles should govern the degree of these interactive forces which has been clearly observed in a clay-polyampholyte system earlier [25]. In the [9] model these specific issues are addressed in finer details. They use a two potential model (repulsive excluded volume interaction and attractive bonding interactions) to develop a unified gelation theory. This model further includes the effect of bond fluctuation dynamics within random percolation framework. The critical exponent values obtained are mostly consistent with the same predicted by this model. All the exponents are compiled and compared in Table 6.2.

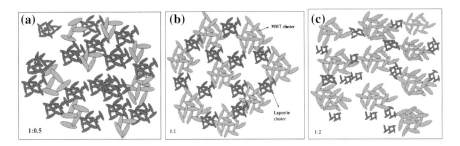

Fig. 6.14 Models of Laponite-MMT assembled network structures shown for various mixing ratios. Note that 1:1 cogel (**a**) has homogeneously structured networks that gives superior mechanical properties unlike the other two cogels (**b** and **c**) where significant amount of heterogeneity is seen. Large and small ellipses represent MMT and Laponite clay particles respectively

6.4 Conclusion

We undertook a detailed study on discotic sols and gels, comprising of Laponite and MMT, in order to understand the associated sol and gel phase behaviour during pre and post gelation scenario. The sol state exhibited zeta potential and gelation concentration values that could be expressed as linear combination of the same of their constituents. Clay cogels prepared with varying mixing ratios showed enhanced mechanical and thermal properties. The mixed discotic colloidal systems were found to be having gel-like features: well defined gelation concentration and finite rigidity modulus. In addition, the experimental data, in hand, mostly supported percolation type gelation mechanism. The hyperscaling exponent could not be retrieved from experimental data which has raised some cause for concern. A representative depiction of percolation-type assembly of the clay particles was shown in Fig. 6.14(a–b). Regardless of the fact that a large volume of work existing in the literature, the formation of gels in pure Laponite and MMT suspensions remains poorly understood. Thus, the understanding of the dispersed state behavior of the mixed system poses various difficulties. For example, whilst deionized water was used, and no salt was added, no attempt at dialysis was made and we would suggest the ionic strength, whilst low, was not really known. Ingress of CO_2 from the air will also mean ionic strength was not constant. For clay particles at low ionic strength, any ionic impurities will really matter for the suspension properties so not controlling the ionic strength was a problem. At the same time, it was possible to note several universal features both in sol and gel states. The data available permitted us to conclude that an interconnected network like structure was formed as the gelation concentration was approached. Thus, the mixed clay sol transforms to gel like in gelation of polymers. The cogels formed with equal concentration of both clays had high network rigidity (or yield strength) and was homogeneous at room temperature. The gel hardening transition observed in all the clay gels was also studied and modeled, and the results will be presented in the next chapters.

References

1. M. Adam, M. Delsanti, D. Durand, Mechanical measurements in the reaction bath during the polycondensation reaction, near the gelation threshold. Macromolecules **18**, 2285–2290 (1985)
2. A. Aharony, D. Stauffer, *Introduction to Percolation Theory* (Taylor and Francis, London, 1994)
3. A. Ajji, L. Choplin, Rheology and dynamics near phase separation in a polymer blend: model and scaling analysis. Macromolecules **24**, 5221–5223 (1991)
4. F.J. Arroyo, F. Carrique, M.L. Jimenez-Olivares, A.V. Delgado, Rheological and electrokinetic properties of sodium montmorillonite suspensions: II. low-frequency dielectric dispersion. J. Colloid Interface Sci. **229**, 118–122 (2000)
5. H.A. Baghdadi, J. Parrella, R. Bhatia, Evidence for re-entrant behavior in Laponite-PEO systems. Rheol. Acta **47**, 349–357 (2008)
6. P. Bala, B.K. Samantaray, S.K. Srivastava, Dehydration transformation in ca-montmorillonite. Bull. Mater. Sci. **23**, 61–67 (2000)
7. D. Bonn, H. Kellay, H. Tanaka, G. Wegdam, J. Meunier, Laponite: what is the difference between a gel and a glass? Langmuir **15**, 7534–7566 (1999)
8. P. Chen, L. Zhang, Interaction and properties of highly exfoliated soy protein/montmorillonite nanocomposites. Biomacromolecules **7**, 1700–1706 (2006)
9. L. de Arcangelis, E. Del Gado, A. Coniglio, Complex dynamics in gelling systems. Eur. Phys. J. E **9**, 277–282 (2002)
10. P.G. de Gennes, *Scaling Concepts in Polymer Physics* (Cornell University Press, Ithaca, 1979)
11. M. Dikstra, J.P. Hansen, P.A. Madden, Gelation of a clay colloid suspension. Phys. Rev. Lett. **75**, 2236–2239 (1995)
12. J.D.G. Durán, M.M. Ramos-Tejada, F.J. Arroyo, F. González-Caballero, Rheological and electrokinetic properties of sodium montmorillonite suspensions: I. rheological properties and interparticle energy of interaction. J. Colloid Interface Sci. **2000**(229), 107–117 (2000)
13. D. Durand, M. Delsanti, M. Adam, J.M. Luck, Frequency dependence of viscoelastic properties of branched polymers near gelation threshold. Europhys. Lett. **3**, 297 (1987)
14. J.-C.P. Gabriel, C. Sanchez, P. Davidson, Observation of nematic liquid-crystal textures in aqueous gels of smectite clays. J. Phys. Chem. **100**, 11139–11143 (1996)
15. J.W. Gilman, Flammability and thermal stability studies of polymer layered-silicate (clay) nanocomposites. Appl. Clay Sci. **15**, 19–31 (1999)
16. M. Kroon, G.H. Wegdam, R. Sprik, Dynamic light scattering studies on the sol–gel transition of a suspension of anisotropic colloidal particles. Phys. Rev. E **54**, 6541–6550 (1996)
17. J. Labanda, J. Llorens, Influence of sodium polyacrylate on the rheology of aqueous Laponite dispersions. J. Colloid Interface Sci. **289**, 86–93 (2005)
18. Laporte Industie Ltd, Laponite Tech. Bull. L 104/90/A (1990)
19. C. Lu, Y.W. Mai, Influence of aspect ratio on barrier properties of polymer-clay nanocomposites. Phys. Rev. Lett. **95**, 088303 (2005). (1–4)
20. L.J. Michot, I. Bihannic, S. Maddi, S.S. Funari, C. Baravian, P. Levitz, P. Davidson, Liquid-crystalline aqueous clay suspensions. Proc. Natl. Acad. Sci. USA **103**, 16101–16104 (2006)
21. P. Mongondry, J.F. Tassin, T. Nicolai, Revised state diagram of Laponite dispersions. J. Colloid Interface Sci. **283**, 397–405 (2005)
22. A. Mourchid, P. Levitz, Long-term gelation of Laponite aqueous dispersions. Phys. Rev. E **1998**(57), 4887–4890 (1998)
23. A. Mourchid, A. Delville, J. Lambard, E. LeColier, P. Levitz, Phase diagram of colloidal dispersions of anisotropic charged particles: equilibrium properties, structure, and rheology of laponite suspensions. Langmuir **11**, 1942–1950 (1995)
24. T. Nicolai, S. Cocard, Light scattering study of the dispersion of Laponite. Langmuir **16**, 8189–8193 (2000)

25. N. Pawar, and H. B. Bohidar, 2009. Surface selective binding of nanoclay particles to polyampholyte protein chains. J. Chem. Phys. 131:045103(1-10)
26. F. Pignon, J.M. Piau, A. Magnin, Structure and pertinent length scale of a discotic clay gel. Phys. Rev. Lett. **76**, 4857–4860 (1996)
27. F. Pignon, A. Magnin, J.M. Piau, B. Cabane, P. Lindner, O. Diat, Yield stress thixotropic clay suspension: Investigations of structure by light, neutron, and X-ray scattering. Phys. Rev. E **56**, 3281–3289 (1997)
28. B. Ruzicka, E. Zaccarelli, A fresh look at the Laponite phase diagram. Soft Matter **7**, 1268–1286 (2011)
29. B. Ruzicka, L. Zulian, G. Ruocco, More on the phase diagram of Laponite. Langmuir **22**, 1106–1111 (2006)
30. B. Ruzicka, L. Zulian, R. Angelini, M. Sztucki, A. Moussaid, G. Ruocco, Arrested state of clay-water suspensions: gel or glass? Phys. Rev. E **77**, 020402 (2008). (1–4)
31. B. Ruzicka, L. Zulian, E. Zaccarelli, R. Angelini, M. Sztucki, A. Moussaid, G. Ruocco, Competing interactions in arrested states of colloidal clays. Phys. Rev. Lett. **104**, 085701 (2010). (1-4)
32. A. Shalkevich, A. Stradner, S.K. Bhat, F. Muller, P. Schurtenberger, Cluster, glass, and gel formation and viscoelastic phase separation in aqueous clay suspensions. Langmuir **23**, 3570–3580 (2007)
33. D. Stauffer, A. Coniglio, A. Adams, Polymer networks. Adv. Polym. Sci. **44**, 103–158 (1982)
34. H. Tanaka, J. Meunier, D. Bonn, Nonergodic states of charged colloidal suspensions: repulsive and attractive glasses and gels. Phys. Rev. E **69**, 031404 (2004)
35. H. Van Olphen, *An Introduction to Clay Colloid Chemistry* (Willey and Sons, New York, 1997)

Chapter 7
Aging Dynamics in Mixed Nanoclay Dispersions

Abstract This chapter investigates the ergodicity breaking and aging dynamics in mixture of colloidal clays, Laponite (L) and Montmorillonite (MMT). The relaxation dynamics has been studied systematically through the light scattering experiments. Spontaneously evolved phases like gel and glass phases have been identified. The growth of anisotropy of the system with age of the sample is observed in the mixed clay dispersions for the first time ever in the literature.

7.1 Introduction

Though considerable effort has been made in the past to offer a systematic and comprehensive explanation to the observed aging of arrested phase of Laponite dispersions, very few studies have been undertaken to probe the same in mixed clay systems which constitutes the objective of this chapter. We have used the aqueous dispersions of popular clays Laponite (L) (aspect ratio ≈30) and Montmorillonite (MMT) (aspect ratio ≈200) mixed in 1:1 weight ratio, and systematically studied their aging dynamics in the clay concentration range, $1.5 < c < 3$ % (w/v). Both these clays are well known in the literature as far as their physical attributes and dispersion phase characteristics are concerned and the phase diagrams of both the clays were investigated in Chaps. 3 and 5. This raises the pertinent question: How does the arrested phase of the mixed system evolve and age? We addressed these issues in this chapter. In our earlier studies, phase behavior of L-MMT clays during their pre and post gelation scenario was discussed [19, 20]. Thus, the anisotropy associated with platelet charge and structure give rise to peculiar dynamical properties, which are exploited in processes like drilling and in customizing products such as cement, paper, paint and composite materials [7, 23].

7.2 Sample Preparation

The samples were prepared as described in Chap. 6. Several sets of required L-MMT mixed dispersions were prepared by mixing the individual suspensions in 1:1 volumetric ratio and stirring the same for 10 min to generate optically clear and homogeneous dispersions. The stopping point of stirring was considered as $t_w = 0$. Samples were stored in air tight borosilicate glass vials when not in use. It was observed that in the literature both volume fraction and % (w/v) units have been used to designate clay concentration. We have used volume fraction to represent solid concentration mostly, but at places % (w/v) unit was be used to enable comparison with literature data.

7.3 Results and Discussion

7.3.1 Concentration Dependence

At the outset it was felt imperative to probe the effect of concentration ϕ on spatial arrangement of platelets in various dispersions in the Guinier regime, $qR_g \ll 1$, R_g being the radius of gyration of the scattering moiety. This was achieved through the measurement of intensity of light $I(q, \phi)$ scattered off these samples in the $q \to 0$ region and the data is presented in Fig. 7.1. These results reveal two regimes for the Laponite and mixed dispersions, but a single regime for the MMT samples.

We observed that the variation in the intensity of the light scattered off Laponite and mixed dispersions with volume fraction up to $\phi = 0.0079$ followed a power law, $I(q, \phi) \sim \phi^{\alpha}$, where $\alpha = 0.95$ and 1.49 for Laponite and mixed system respectively, and remained constant for $\phi > 0.0079$. But, for MMT samples, $\alpha = 1.38$ and no saturation was noticed. The cut off observed in both Laponite and L-MMT dispersions was same, $\phi_{\text{cutoff}} = 0.0079$. A weak dependence on ϕ was detected above this cut off, which may be attributed to the random arrangement of platelets in the system. These measurements were done at the initial time $t_w = 0$. The theory of scattering predicts that a dispersion containing a distribution of clusters given by cluster size distribution n_i would follow [15]

$$\langle I(q, \phi) \rangle = \langle I_0 \rangle \sum_i n_i i^2 \qquad (7.1)$$

where i is the ratio of cluster to monomer molecular weight and defines the aggregation number of the cluster and $\langle I_0 \rangle$ is the reference intensity. Thus, measured scattered intensity is insensitive to the geometrical structure of scattering centers in the $qR_g \ll 1$ regime, but it is very sensitive to their physical size. Thus, the growth of intensity could be attributed to either association of platelets or to excluded volume interactions. Milton [15] through

7.3 Results and Discussion

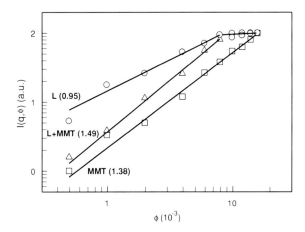

Fig. 7.1 The variation of intensity of light scattered from the dispersions of Laponite, Laponite-MMT and MMT measured at room temperature. The power-law exponents (α) are indicated in the figure for each *curve*. *Solid lines* are least-squares fitting of data to power-law: $I(q, t) \sim \phi^{\alpha}$

various model calculations applicable for hard sphere dispersions has shown that $\langle I(q,\phi) \rangle$ varies with solute concentration as: (1) when full accounting of excluded volume interactions is considered intensity grows linearly in the low concentration region ($c < 2$ % (w/v)) followed by a slow variation domain when $c > 5$ % (w/v) and (2) when no such correction is accounted for the said dependence is almost linear in the entire concentration region. Thus, it could be argued that the platelets of Laponite, MMT and mixed system are associating in an excluded volume environment. Baravian et al. [1] have found that the shear thinning of flow in dilute and semi dilute clay dispersions with $\phi < 0.0025$ in low ionic strength conditions (<5 mM) can be explained on the basis of excluded volume effects [1, 14]. It was not possible to infer anything more from this data at this stage. A detailed study devoted to aging dynamics is the main focus of this chapter.

7.3.2 Ergodicity Breaking Time

The aging behavior and slow dynamics is best probed by dynamic light scattering experiments. The dynamic structure factors were obtained as described in Chap. 2 for the non-ergodic sample and were plotted in the Figs. 7.2 and 7.3.

Figure 7.2 showed the temporal evolution of $g_1(q, t)$ for, the sample ($\phi = 0.00988$) above $\phi_{cutoff} = 0.0079$. Each plot showed similar temporal evolution dynamics. It was clearly seen that $g_1(q, t)$ did not completely relax as the samples age indicating confinement of platelets in the arrested phase. Thus, as the sample aged the decay of the relaxations became slower, meaning that platelets had reduced mobility that gave rise to less rapid intensity fluctuations.

The data shown in Fig. 7.2 could not be fitted either to a single exponential or to a single stretched exponential function with acceptable statistical accuracy. Best least-squares fit were obtained when the data was fitted to two different relaxation

Fig. 7.2 The evolution of the dynamic structure factor of the mixed system shown for sample of volume fraction: $\phi_{\text{cutoff}} < \phi = 0.00988$ (2.5 % (w/v)). The *solid curves* represent fitting of the data to Eq. 7.2. The *arrows* indicate the evolution of structure factor starting $t_w = 0$; these data was taken every 500s

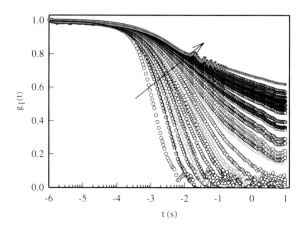

Fig. 7.3 The evolution of the dynamic structure factor of the mixed system shown for samples of volume fractions: **a** $\phi_{\text{cutoff}} > \phi = 0.0059$ [1.5 % (w/v)] **b** $\phi_{\text{cutoff}} = 0.0079$ [2.0 % (w/v)]. The *solid curves* represent fitting of the data to Eq. 7.2. The *arrows* indicate the evolution of structure factor starting $t_w = 0$; these data were taken every 700 s and 1000 s for $\phi = 0.0079$ and 0.0059 respectively

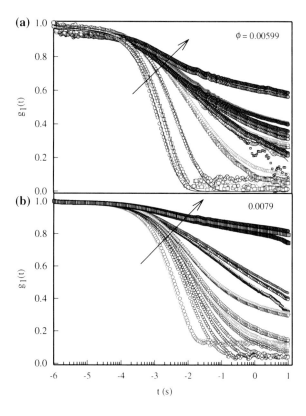

processes: a fast and a slow one. Therefore the fitting expression should contain two contributions. A practical approach to describe the shape of the experimental autocorrelation function is the sum of exponential functions given by Eq. 7.2.

7.3 Results and Discussion

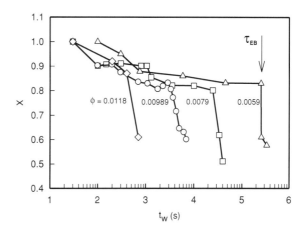

Fig. 7.4 Semi-log plot of the heterodyne contribution parameter X shown as function of aging time, t_w. The point where there is sharp decrease in the value of X is defined as the ergodicity breaking time, τ_{EB}, see the *arrow*. *Solid lines* are guide to the eye

Therefore dynamic structure factor $g_1(q,t)$ is described as

$$g_1(q,t) = a\exp-\left(\frac{t}{\tau_1}\right) + (1-a)\exp-\left(\frac{t}{\tau_2}\right)^\beta \qquad (7.2)$$

where a and $(1-a)$ are the weights of the two contributions τ_1 and τ_2, which in turn are the fast and the slow mode relaxation times respectively, and β is the stretching parameter. The fast mode relaxation time τ_1 is related to the inverse of the short-time diffusion coefficient D_s as $\tau_1 = 1/D_s q^2$. The stretched exponential function has been reported in the past for clay dispersions, since it has been found empirically that it provides good description of the slow relaxation processes encountered in arrested systems; this would be discussed later.

Figure 7.4 shows the variation of the heterodyne contribution, X, with the aging time t_w, and the ergodicity breaking point is defined as the time where there is a sharp change in the value of X parameter.

As the clay concentration was increased the system developed non-ergodic phase rather quickly. Interestingly, the ergodicity breaking time obtained from experimental data decreased exponentially with the solid concentration as shown in the Fig. 7.5. Consequently, an empirical Eq. 7.3 could be proposed

$$\tau_{EB} \sim t_0 \exp(-\phi/\phi_0) \qquad (7.3)$$

where $t_0 = 2.55 \times 10^8$ s, and $\phi_0 = 8.6 \times 10^{-4}$. Equation 7.3 defines an observation that was not reported hitherto in any clay dispersion.

7.3.3 Relaxation Dynamics

The collective dynamics operative in mixed clay dispersions above the volume fraction ϕ_{cutoff} was probed using dynamic light scattering studies. Normally a scattering system like clay glass has a complicated free-energy landscape with many

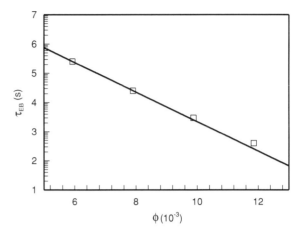

Fig. 7.5 Semi-log plot of clay concentration on the ergodicity breaking time shown for the mixed clay dispersion. Notice that non-ergodicity was achieved earlier in concentrated samples

local minima. Such a system is associated with a relaxation time distribution which yields a multi-modal correlation function $g_1(q, t)$. The Laplace inverse transform of the dynamic structure factor, $g_1(q, t)$, $G(q, \tau) = \int_{\tau_{min}}^{\tau_{max}} g_1(q,t) \exp(t/\tau) dt$ yields the relaxation time distribution function, $G(q, t)$ [3]. We focused on the behavior of relaxation times of the mixed system at different solid contents from here on. Figure 7.6 showed the distribution of relaxation times as function of waiting time for various samples.

It was clear from Fig. 7.6 that there existed two distinct modes of relaxations in the Laponite and mixed system, whereas only a single relaxation exists in the MMT dispersion. The arrested phase in Laponite dispersion has been reported to contain two relaxation modes [2, 8, 9]. Furthermore, from studying the angular dependence of the scattering function $g_1(q, t)$, it was found that the fast mode relaxation time τ_1 varied inversely with q^2 shown in Fig. 7.7. Therefore, this fast mode was found to be diffusive and independent of aging time. This implied that for the short times, the particles undergo normal Brownian motion [3]. For Laponite system invariance of the diffusive fast mode relaxation with aging time has been reported earlier [2, 8, 9]. This study confirmed the same for the L-MMT samples.

As was the case for τ_1, the relaxation time τ_2 was found to scale with the q^{-2}, as shown in Fig. 7.7 feature again reminiscent of classical diffusion for times less than ergodicity breaking time, although the stretched exponential relaxation of slow mode shows that there are strong hydrodynamic interactions between particles. After the ergodicity breaking time, the motions of the particles is no more diffusive and begin to scale as $\tau_2 \sim q^{-1}$. Bellour et al. [2] used multispeckle technique to account for non-ergodicity in Laponite glass and found identical spatial dependence of slow mode relaxation time. Kaloun et al. [10] observed similar dependence in an experiment where the dynamics of tracer diffusion was probed inside Laponite glass in the linear aging regime.

Kaloun et al. [10] and others [2, 8, 9] reported two aging regimes in colloidal Laponite glass: in the first regime, slow mode relaxation time, τ_2 was observed to

7.3 Results and Discussion

Fig. 7.6 Plot shows the distribution of relaxation time **a** for Laponite suspension of $\phi = 0.00989$ at different aging time and **b** for the Laponite ($\phi = 0.00989$), L-MMT ($\phi = 0.00989$) and MMT ($\phi = 0.00874$) at the initial time $t_w = 0$. Plot shows the distribution of relaxation time evolution for Laponite-MMT suspension of $\phi = 0.00989$ at different aging times. Note the relative invariance of fast mode and broadening of the slow mode for aged samples

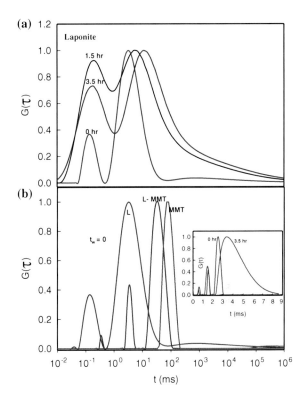

Fig. 7.7 The characteristic times, τ_1 and τ_2 of the fast and slow relaxations for L-MMT sample, shown as function of the wave vector q, for a $\phi = 0.00988$ ($= 2.5\%$ w/v) sample. The *top* plot is in the sol state before ergodicity breaking and evolves to *bottom* plot after ergodicity breaking due to aging

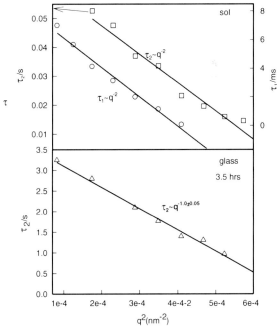

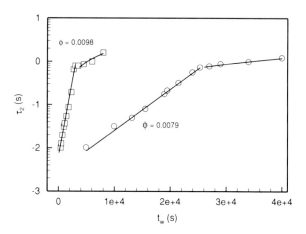

Fig. 7.8 Slow mode relaxation time shown as function of the aging time for $\phi = 0.0079$ (*circle*) and 0.00989 (*square*) for L-MMT suspensions determined from dynamic structure factor data

increase exponentially with t_w, followed by a full aging region where τ_2 was proportional to t_w. The aging behavior in L-MMT glass was examined and the temporal evolution of slow mode relaxation time τ_2 is presented in Fig. 7.8 before the ergodicity breaking time. The data allows the relaxation time τ_2 to be characterized as

$$\tau_2 \sim \tau_0 \exp(t_w/t_0) \qquad (7.4)$$

where $\tau_0 \sim 2.8$ ms, $t_0 \sim 0.21$ ms for $\phi = 0.0079$ and $\tau_0 \sim 6.47$ ms, $t_0 \sim 0.16$ ms for $\phi = 0.00989$.

The temporal evolution of slow mode relaxation time τ_2 in full aging regime i.e. above τ_{EB}, follows a power law growth as

$$\tau_2 \sim t_w^y \qquad (7.5)$$

where $y = 1.1 \pm 0.02$ for $\phi = 0.0079$ and $y = 0.97 \pm 0.02$ for $\phi = 0.00989$.

Thus, it was interesting to note that the arrested phase of the L-MMT system aged alike the Laponite glass regardless of the presence of MMT. This feature was seen during the initial aging, $t_w < 10^4$ s. For the mixed system and $\phi > \phi_{cutoff}$, the short time aging prevailed in the range $t_w < 2 \times 10^3$ whereas for $\phi \approx \phi_{cutoff}$ the same was observed in the time span $t_w < 3 \times 10^4$. Thus, a concentrated system aged much faster.

The slow mode decay time of the correlation function is a measure of the time a particle needs to forget its initial position, these results show that very rapidly the aging freezes in a certain degree of the system. However, the low and high concentration samples were seen to exhibit distinctive behavior as far as dynamic arrest is concerned, a phenomenon that was repeatedly noted in several measured parameters.

Finally, the stretched exponent or the width parameter was found to depend on the aging time. The exponent β was found to decrease linearly between 1 and 0, as shown in Fig. 7.9. The aging time t_w for which β has the lowest value corresponds to the ergodic-non ergodic transition on the laboratory time scale. It was observed

7.3 Results and Discussion

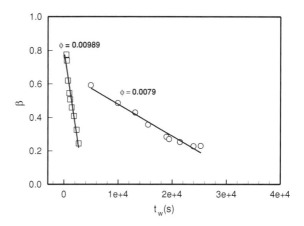

Fig. 7.9 Variation of stretch exponent β of the slow relaxation mode as a function of the aging time for L-MMT suspensions. *Solid lines* are least-squares fitting of data to *straight lines*

that both the $\phi \approx \phi_{\text{cutoff}}$ and $\phi > \phi_{\text{cutoff}}$ concentration samples were associated with β value close to 0.2 at late times and at $t_w = 0$, both started with same ergodicity, $\beta \approx 0.7 \pm 0.1$. Jabbari-Farouji et al. [9] observed similar results for Laponite glass.

7.3.4 Behavior of the System $\phi < \phi_{cutoff}$

It was felt imperative to examine this behaviour for dilute systems, $\phi < \phi_{cutoff}$. The representative data is shown in Fig. 7.10. The dynamics for $\phi = 0.0059$ was found to be quite different from the dispersions at higher concentrations. The slow mode relaxation time τ_2 was seen to grow much faster than exponential function and stretch exponent β was also not linearly dependent on t_w, which clearly revealed that the low concentration dynamics was different, which was not fully explored in the current studies.

In a study by Jabbari-Farouji et al. [8] it was reported that Laponite system undergoes a hesitation between formation of gels and glasses for the range of concentrations $1.4 < c < 2.3$ %, but we did not observe such behavior for the same concentrations in our mixed system. The reproducibility of the data was verified by repeating the experiments 4 times. Our mixed system showed clear regions of gel ($c < 2$ % w/v) and glass ($c > 2$ % w/v) states.

7.3.5 Growth of Anisotropy with Aging

It has been reported in the literature that Laponite clay glasses exhibit anisotropy with aging. In an interesting study, Shahin et al. [22] have shown that the anisotropy propagates downwards with time starting from the interface. Kinetics of the

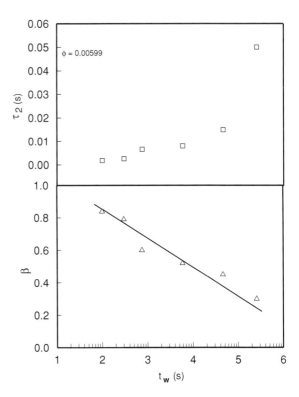

Fig. 7.10 Characteristic time τ_2 and the stretch exponent β of the slow mode relaxation as function of the aging time for the dilute L-MMT suspension, $\phi = 0.0059$ ($\phi < \phi_{\text{cutoff}}$, corresponds 1.5 % w/v)

orientational ordering was studied in Chap. 4. The aging dependent growth of anisotropy in the mixed glass was examined through the measurement of depolarized component of light scattered from these samples.

Depolarization ratio is defined as follows [3]

$$D_p = \frac{I_{VH}}{I_{VV}} \qquad (7.6)$$

where I_{VH} and I_{VV} are the depolarized and polarized components of the light scattered by a medium in the plane parallel and perpendicular to the incident linearly polarized laser beam. Depolarization ratio is a quantitative measure of spatial anisotropy present in the scattering medium. Figure 7.11 depicted the growth in anisotropy with waiting time for L + MMT glass, for comparison the data from Laponite and MMT samples were shown in the same figure. The data reveals two distinct signatures: neither Laponite nor MMT dispersions showed much anisotropy whereas the L + MMT glass displayed considerable depolarized scattering during aging. This growth could be described by a power-law expression

$$D_p \sim t_w^{0.46} \qquad (7.7)$$

7.3 Results and Discussion

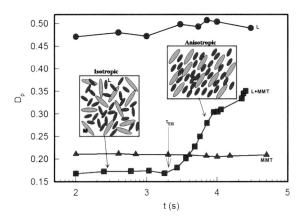

Fig. 7.11 Depolarization ratio of the Laponite (L), Laponite + MMT ($\phi = 0.00989$, L + MMT) and MMT suspensions as function of the aging time t_w. Note the sharp increase in D_p at $t_w = \tau_{EB}$. Schematic illustration of the arrangement of clay particles are shown at different aging times

Interestingly, the initiation of rapid growth in anisotropy exactly matches with the ergodicity breaking time discussed earlier. Laponite dispersions also show a marginal increase in the anisotropy, but it is almost constant in the experimental time as shown in the Fig. 7.11. We examined a Laponite sample over a period of 20 days, and noticed very slow growth with $D_p \sim t_w^{0.15}$. It was observed that the MMT particles present in the system tried to induce isotropy and the Laponite forced the system towards anisotropy. There are few studies in the literature on the anisotropy of Laponite system [12, 13, 16, 18, 22]. This work was the first study on the mixed clay system where the anisotropy of the system was conclusively shown to grow exactly at the ergodicity breaking point.

7.3.6 Cole–Cole Plot and Sample Heterogeneity

The phase homogeneity in polymer solutions and melts is often deduced from Cole–Cole plot where the imaginary part of the complex viscosity (η'') is plotted as function of the real part (η'). Typically, in a melt, at very low frequency, viscous behaviour is observed whereas at higher frequencies elastic properties dominate. In this formalism, $\eta^*(\omega) = \eta'(\omega) + i\eta''(\omega)$ and the low and high frequency viscosity values are given by η_0 and η_∞ respectively. The Cole–Cole empirical expression is written as [6]

$$\eta^* - \eta_\infty = \frac{(\eta_0 - \eta_\infty)}{\left[1 + (j\omega\tau_{cc})^{1-\alpha}\right]}; \quad 0 < \alpha < 1 \tag{7.8}$$

The aforesaid expression is interpreted as arising from a superposition of several Debye relaxations [5, 24]. The mean relaxation time is given by τ_{cc}. This plot has been used extensively to map homogeneity of soft matter systems and their composites. For a homogeneous phase, the Cole–Cole plot is a perfect semicircle

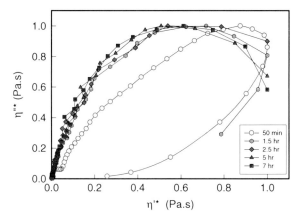

Fig. 7.12 Cole–Cole plot of the Complex for $\phi = 0.00989$ ($c = 2.5$ % (w/v)), which is plotted for different waiting times, t_w where $\eta'^* = \eta'/\eta'_{max}$ and $\eta''^* = \eta''/\eta''_{max}$

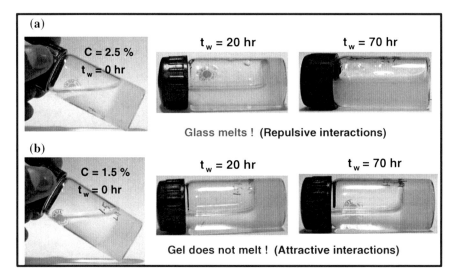

Fig. 7.13 Dilution experiments are done by adding equal amount of water to the Complex. **a** Represents the Wigner glass, where repulsive interactions are present and **b** the gel where attractive interactions dominate

($\alpha = 0$) with a well defined relaxation time. Any deviation from this shape indicates non-homogeneous dispersion and phase segregation due to immiscibility. Such phases are associated with relaxation time distributions mentioned earlier.

The Cole–Cole plot initially ($t_w = 50$ min) was seen to be associated with a smaller slope in the low viscosity regime and an elliptical contour indicating the presence of heterogeneity in the arrested phase as shown in Fig. 7.12. As the waiting time grew, the initial slope of the plots were found to increase considerably, but largely remained invariant of t_w. Such behaviour clearly refers to enhanced heterogeneity in the sample. Thus, the mixed glass of L + MMT became anisotropic and heterogeneous simultaneously with aging.

7.4 Dilution Experiment

Dilution experiments were used to identify the gel and glass phases of the system as shown in Fig. 7.13. For the concentrations $\phi > \phi_{cutoff}$, melting of the samples was observed by adding the similar solvent, which gave the indication of dominance of repulsive interactions. Attractive interactions dominated in the samples $\phi < \phi_{cutoff}$ which confirmed the gel state. Thus the dilution test gave the clear indication of different states of the system.

7.5 Conclusion

We systematically probed the slow dynamics in aging Laponite-MMT glass system at room temperature. In order to confirm that the arrested phase was indeed a glass phase, dilution tests were carried out on these samples following Ruzicka et al. [21] prescription which established the presence of glass phase in our system. Interestingly, the two key characteristic parameters defining the aging of the system like, the fast and slow mode relaxation time of the dynamic structure factor, and the concentration dependence of ergodicity breaking time, exactly followed the trend observed in Laponite glass. It was observed that the ergodicity breaking time was an exponentially decaying function of clay concentration alike what is seen in Laponite glass. Similarly, the observed cross-over in the spatial dependence of slow mode relaxation time from q^{-2} to q^{-1} with aging was earlier seen in Laponite system. Thus, it could be unambiguously concluded that the aging pathway observed in Laponite-MMT glass followed qualitatively same footprints of Laponite glass. It has been reported that MMT platelets are associated with a disc diameter, $d \approx 200$ nm and zeta potential, $\zeta \approx -30$ mV whereas the same for Laponite particles is $d = 20$ nm and $\zeta = -52$ mV respectively [11]. Zeta potential is linearly dependent of net surface charge to a very good approximation [17]. Thus, typically, the surface charge density of a particle is $\approx (\zeta/d^2)$ which indicates that Laponite charge density is 100-times higher than that of MMT. Consequently, all the inter-platelet interactions will be primarily dominated and governed by Laponite–Laponite electrostatic forces in the mean field provided by the weakly charged MMT platelets. This attributes a matrix-like platform provided by MMT on which the Laponite particles strongly interact. Such a description is consistent with the experimental observations. The second aging regime reported in the literature [4, 10] was not observed in our studies may be because our upper limit of t_w was not long enough. Further experiments carried out in the present work identify the said phases as glass-like. The hypothesis of Laponite platelets interacting in a mean-field environment provided by the MMT matrix needs further exploration.

References

1. C. Baravian, D. Vantelon, F. Thomas, Rheological determination of interaction potential energy for aqueous clay suspensions. Langmuir **19**, 8109 (2003)
2. M. Bellour, A. Knaebel, J.L. Harden, F. Lequeux, J.-P. Munch, Aging processes and scale dependence in soft glassy colloidal suspensions. Phys. Rev. E **67**, 031405 (2003)
3. B.J. Berne, R. Pecora, Dynamic light scatteing with applications to chemistry, biology and physics. (Wiley-Interscience, New York, 1976)
4. L. Cipelletti, S. Manley, R.C. Ball, D.A. Weitz, Universal aging features in the restructuring of fractal colloidal gels. Phys. Rev. Lett. **84**, 2275 (2000)
5. K.S. Cole, R.H. Cole, Dispersion and absorption in dielectrics I alternating current characteristics. J. Chem. Phys. **9**, 341–351 (1941)
6. D.W. Davidson, Dielectric relaxation in liquids: I. the presentation of relaxation behavior. Can. J. Chem. **39**, 571–594 (1961)
7. K. Faisandier, C.H. Pons, D. Tchoubar, F. Thomas, Structural organization of Na- and K-montmorillonite suspensions response to osmotic and thermal stresses. Clay Clay Miner. **46**, 636 (1998)
8. S. Jabbari-Farouji, E. Eiser, G.H. Wegdam, D. Bonn, Ageing dynamics of translational and rotational diffusion in a colloidal glass. J. Phys.: Condens. Matter **16**, 471 (2004)
9. S. Jabbari-Farouji, G.H. Wegdam, D. Bonn, Gels and glasses in a single system: evidence for an intricate free-energy landscape of glassy materials. Phys. Rev. Lett. **99**, 065701–065704 (2007)
10. S. Kaloun, R. Skouri, M. Skouri, J.P. Munch, F. Schosseler, Successive exponential and full aging regimes evidenced by tracer diffusion in a colloidal glass. Phys. Rev. E **72**, 011403 (2005)
11. J. Labanda, J. Llorens, Influence of sodium polyacrylate on the rheology of aqueous Laponite dispersions. J. Colloid Interface Sci. **289**, 86–93 (2005)
12. B.J. Lemaire, P. Panine, J.C.P. Gabriel, P. Davidson, The measurement by SAXS of the nematic order parameter of Laponite gels. Europhys. Lett. **59**, 55 (2002)
13. C. Martin, F. Pignon, A. Magnin, M. Meireles, V. LeliÃvre, P. Lindner, B. Cabane, Osmotic compression and expansion of highly ordered clay dispersions. Langmuir **22**, 4065 (2006)
14. L.J. Michot, C. Baravian, I. Bihhanic, S. Maddi, C. Moyne, J.F.L. Duval, P. Levitz, P. Davidson, Sol–gel and isotropic/nematic transitions in aqueous suspensions of natural nontronite clay. influence of particle anisotropy. 2. Gel Struct. Mech. Prop. Langmuir **25**, 127–139 (2009)
15. A.P. Minton, Static light scattering from concentrated protein solutions, I: general theory for protein mixtures and application to self-associating proteins. Biophys. J. **93**, 1321 (2007)
16. A. Mourchid, A. Delville, J. Lambard, E. LeColier, P. Levitz, Phase diagram of colloidal dispersions of anisotropic charged particles: equilibrium properties, structure, and rheology of Laponite suspensions. Langmuir **11**, 1942–1950 (1995)
17. H. Oshima, Electrophoresis of soft particles. Adv. Colloid Interface Sci. **62**, 189 (1995)
18. F. Pignon, M. Abyan, C. David, A. Magnin, M. Sztucki, In situ characterization by SAXS of concentration polarization layers during cross-flow ultrafiltration of laponite dispersions. Langmuir **28**, 1083–1094 (2012)
19. R.K. Pujala, N. Pawar, H.B. Bohidar, Universal sol state behavior and gelation kinetics in mixed clay dispersions. Langmuir **27**, 5193 (2011)
20. R.K. Pujala, N. Pawar, H.B. Bohidar, Landau theory description of observed isotropic to anisotropic phase transition in mixed clay gels. J. Chem. Phys. **134**, 194904 (2011)
21. B. Ruzicka, L. Zulian, E. Zaccarelli, R. Angelini, M. Sztucki, A. Moussaïd, G. Ruocco, Competing interactions in arrested states of colloidal clays. Phys. Rev. Lett. **104**, 085701–085704 (2010)

22. A. Shahin, Y.M. Joshi, S.A. Ramakrishna, Interface-induced anisotropy and the nematic glass/gel state in jammed aqueous laponite suspensions. Langmuir **27**, 14045 (2011)
23. H. van Olphen, *An Introduction to Clay Colloid Chemistry* (Wiley, New York, 1997)
24. T.C. Warren, J.L. Schrag, J.D. Ferry, Infinite-dilution viscoelastic properties of poly-γ-benzyl-L-glutamate in helicogenic solvents. Biopolymers **12**, 1905–9015 (1973)

Chapter 8
Thermal Ordering in Mixed Nanoclay Dispersions

Abstract This chapter investigates the temperature induced orientational ordering in 1:1 Laponite-MMT dispersion. This dispersion showed thermally activated irreversible conformational phase transition from mostly isotropic (disordered) to strongly anisotropic (ordered) phase at temperature $\approx$60 °C. This phase transition was modeled through Landau formalism.

8.1 Introduction

This chapter investigates the temperature induced phase transition in mixed clay dispersions. We have studied the aging dynamics of the same system in Chap. 7, where we observed the ordering of the clay particles as they aged towards the arrested phase. This raised a pertinent question: How does thermal dehydration affect the dispersions made of Laponite and MMT clays? We address this specific issue in this chapter. The experimental observations were placed on a sound theoretical footing by modeling the system within the framework of Landau theory of phase transition. The great virtue of Landau theory is that it makes specific predictions for what kind of non-analytic behavior one should see when the underlying free energy is analytic. Then, all the non-analyticity at the critical point, the critical exponents, are because the equilibrium value of the order parameter changes non-analytically, as a square root, whenever the free energy loses its unique minimum.

8.2 Results and Discussion

Samples were prepared as described in Chap. 6. The dispersions used in this study were mixed clays of equal mixing ratio (2.5 % (w/v)) as observed in Chap. 7. The dispersions entered the glass state eventually with time as reported earlier. Enough time (2-days) was given to the samples to stabilize in the arrested phase. The

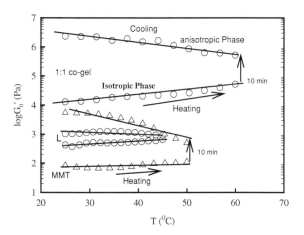

Fig. 8.1 Thermal evolution of dehydration transition in 1:1 co-dispersion of Laponite and MMT. A sharp transition is seen at $T_c \approx 60\,°C$ where the isotropic phase makes an abrupt conformational transition to a dehydrated phase. This arrested phase gains network rigidity during the cooling cycle. *Solid lines are guide to the eye*

influence of temperature on the 1:1 dispersions was studied by a heating-cooling cycle experiment and the data is shown in Fig. 8.1. As the temperature was increased from the ambient (25 °C) to ≈60 °C a dehydration transition was observed. This transition temperature was located at 50 °C for MMT and 45 °C for Laponite samples. Regardless, all the samples exhibited similar hysteresis loop. The storage modulus abruptly increased many fold at the transition temperature (close to T_c) and upon temperature reversal it remained arrested in the same phase. This arrested phase gained further network rigidity during the cooling cycle implying the cooperative nature of solvent loss. This was also indicative of the existence of several free-energy driven morphological states of the colloidal glass. This transition was thermally activated and was observed to be irreversible. This would be discussed further in the light of Landau theory later. Another pertinent observation was made related to the value of storage modulii of various dispersions at room temperature: MMT dispersions were associated with the smallest G'_0 value (~200 Pa), Laponite system has $G'_0 \sim 500$ Pa whereas the mixed dispersion exhibited G'_0 as high as 1,000 Pa which cannot be attributed to any network held with weak interactions. Thus, it is argued that there must be strong electrostatic binding induced network formation even at room temperature in salt free dispersions.

The area under the heating–cooling curve represented the total gain in dispersion strength due to loss of solvent and network reorganization which was found to be considerably large compared to that of individual clay dispersions. The gain in rigidity during the cooling cycle, seen in all samples, implied that the dehydration was a cooperative process.

The dehydration transition to a predominantly anisotropic phase was confirmed from depolarized light scattering data. Figure 8.2 shows the variation of depolarization ratio of light scattered from this dispersion at room temperature. Depolarization ratio [2] is defined as follows

$$D_p = \frac{I_{VH}}{I_{iso}} \qquad (8.1)$$

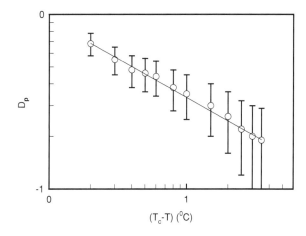

Fig. 8.2 Log–log plot of variation of normalized depolarization ratio for the complex (formed with 1:1 as mixing ratio of two clays) with reduced temperature. *Solid lines* are fitting to the data following Eq. 8.3 (gives $\beta = 0.40 \pm 0.05$). See text for details

and

$$I_{iso} = I_{VV} - \frac{4}{3} I_{VH} \qquad (8.2)$$

where I_{VH} and I_{VV} are the depolarized and polarized components of the scattered intensity. Depolarization ratio is a measure of anisotropy present in the system. In the co-dispersion system the depolarization ratio was observed to increase with increase in temperature up to 60 °C and reached a plateau afterwards (Fig. 8.2). Data presented in Fig. 8.2 implied that anisotropic domains grow preferentially inside the colloidal dispersion with rise in temperature. Initially these were distributed at random. Preferential alignment of these domains was reached at temperature 60 °C and further rise in temperature did not have any impact on these ordered domains. The growth in D_p could be described as

$$D_P \sim (T_c - T)^{-\beta} \qquad (8.3)$$

The exponent was $\beta = 0.40 \pm 0.05$. For MMT dispersions this was 0.1 while for Laponite dispersions the measurements were erratic. This data clearly established that at room temperature the co-dispersion phase was largely isotropic with $D_p \approx 20\%$ which increased to $\approx 70\%$ as temperature reached T_c. This change in anisotropy was gradual below T_c, but abrupt at T_c. The gain in anisotropy owes its origin to the loss of inter planar solvent and the resultant realignment of the colloidal network. Bala et al. [1] observed that the loss of solvent in the low temperature region (<100 °C) can be attributed to the dehydration of inter planar space.

In order to probe this transition further, these samples were subjected to XRD studies. The data is depicted in Fig. 8.3 for a range of temperatures. Peak height clearly quantified the morphological changes affected by temperature change. This is plotted in Fig. 8.4 as a function of temperature. The peak intensity (at $2\theta = 7°$) increased with rise in temperature and was maximum at the transition temperature. This observation revealed that system became more structured after crossing the transition temperature. Interestingly, the transition temperature was again located

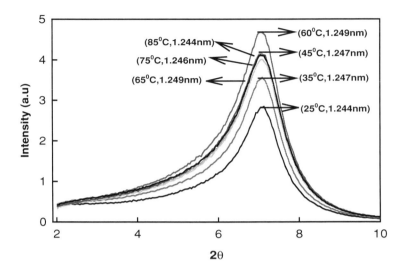

Fig. 8.3 XRD pattern of the 1:1 co-dispersion at different temperature. Temperature and basal plane spacing values are mentioned in the parenthesis for various samples

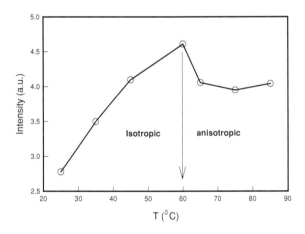

Fig. 8.4 Variation of XRD peak intensity at $2\theta = 7.0$ for the 1:1 co-dispersion for different temperature showing the transition to anisotropic phase. *Solid lines* are guide to the eye. *Arrow* points to transition temperature

at $T_c = 60$ °C. Thus, as far as this transition was concerned it was consistently defined by depolarized light scattering, rheology and XRD data.

Based on the above experimental confirmations, mechanism of phase transition in the co-dispersion system could be modeled. In this model, initially the system was in the hydrated phase with lot of solvent trapped as interstitial water, but when the system was heated to the transition temperature some of this solvent got released as bulk water that forced realignment of the colloidal network thereby

8.2 Results and Discussion

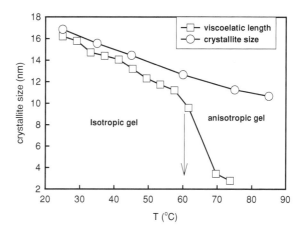

Fig. 8.5 Plot of viscoelastic length and crystallite size as function of temperature

affecting a conformational transition to anisotropic phase. The process was free-energy driven and thermo-irreversible. The anisotropic phase is in a thermodynamically stable state. The XRD data, through Scherrer expression (Eq. 8.4), yields the crystallite size present in the dispersion medium [1].

Neglecting contributions arising from strain and size broadening, the apparent crystallite size, L, is given by Bala et al. [1]

$$L \approx \frac{\lambda_{X-RAY}}{\Delta W \cos\theta} \quad (8.4)$$

where ΔW is full width at half maximum and θ is Bragg scattering angle.

The data is presented in Fig. 8.5. There is a sharp reduction in the crystallite size at T_c which indicates loss of interstitial water. The crystallites correspond to interstitial water domains residing between colloidal discs. Similar observations have been reported earlier [10].

Phase separation kinetics in dynamically asymmetric systems involve stress-diffusion coupling not observed in dynamically symmetric systems. A screening length, called the viscoelastic length ξ, has been introduced to account for the dynamics of viscoelastic relaxations. Physically, a homogeneous network is associated with two characteristic parameters: the growth rate of thermally activated concentration fluctuations, Γ_r and the relaxation rate of entangled networks, Γ_e that is characterized by the viscoelastic properties of the material. Local stress inside the system develops in case the growth rate of concentration fluctuations is faster than the network entanglement relaxation rate, i.e. $\Gamma_r \gg \Gamma_e$. The spatial variation of the stress field will be manifested as an equivalent osmotic pressure field inside the system. This forms the basic formalism of Doi–Onuki model that incorporated the concept of dynamical coupling between stress and diffusion in a system undergoing phase separation [5, 6].

In a network of transiently connected units, the shear modulus is proportional to the concentration of inter-molecular bonds. The value of the length of elastically active strands, ξ, is similar to the characteristic viscoelastic network size, which is estimated from the low-frequency shear modulus, G'_0. This is a measure of elastic free energy stored per unit volume of a characteristic viscoelastic network of size, ξ. In the present case, this parameter estimates the typical separation between colloid rich and poor regions inside the sample at temperature, T and relates to low frequency storage modulus, G'_0, as [5, 6]

$$G'_0 \approx k_B T/\xi^3 \tag{8.5}$$

This equation describes the equivalent of thermal energy stored in the network of volume $\approx \xi^3$ as elastic energy. Thus, typical viscoelastic length scale prevalent in these materials becomes easily accessible from oscillatory rheology measurements. This data is plotted in Fig. 8.5 where a monotonous decrease in this length was observed as the temperature was increased. In the repulsive gel model (Wigner glass) of [11] such a situation would increase repulsive interactions; thereby, increasing the dispersion rigidity.

Though not much work has been reported on binary discotic dispersions, there are some reports on phase transition studies pertaining to Laponite. Ruzicka et al. [8] have observed isotropic liquid to isotropic gel to nematic phase transition in this clay. They attribute this to the aging dynamics prevailing in the system. In a recent work, [10] observed that in the Laponite system, the arrested phase seen at high clay volume fraction is stabilized by screened coulomb interaction (like in in Wigner glass). The computer simulation and SAXS data revealed the long-range character of arrested phase which was attributed to different time-scales controlling the competing attractive and repulsive interactions. In a separate study, [9] provided experimental evidence of the existence and evolution with aging time of two different arrested states in a single system obtained only by changing the volume fraction of the clay. An inhomogeneous phase was observed at low concentration while the high concentration samples yielded a homogeneous state. The ordered phase observed in the current study possibly belongs to a class of arrested phase envisaged by Ruzicka et al. [9]. Ovarlez et al. [7] studied the liquid-solid phase transition in a soft-jammed system and investigated the impact of temperature, density, and load changes on the material behavior. The temperature dependent storage modulus data of MMT dispersions revealed that at higher temperature the dispersion rigidity was enhanced manifold which was attributed to network reorganization. Our results pertaining to storage modulii and gelation temperature are in qualitative agreement with these findings. Neither depolarization data was reported in the literature for clay systems nor was the dehydration transition ever noticed in such systems. In the following discussion, it is argued that the measured depolarization ratio can be used as the control parameter to quantitatively describe the observed dehydration transition behaviour.

8.3 Application of Landau Theory

Landau theory was introduced in an attempt to formulate a general theory of second-order phase transitions where this theory suggests that the free energy of any system should obey two basic conditions: that the free energy is analytic, and that

8.3 Application of Landau Theory

it obeys the symmetry of the Hamiltonian [3]. Given these two conditions, one can write down (in the vicinity of the critical temperature, T_c) a phenomenological expression for the free energy as a Taylor expansion in the order parameter. In the mesoscopic mean-field model conceptualized by Landau, the Gibb's free energy density is described exclusively as function of an order parameter, ψ. Thus, the free energy density is given for a system undergoing phase transition as [3]

$$F(T, \psi) = F_0 + \frac{1}{2}a(T - T^*)\psi^2 + \frac{1}{3}b\psi^3 + \frac{1}{4}c\psi^4 \quad (8.6)$$

where T^* is some characteristic temperature that is not necessarily the transition temperature (T_c), F_0 is the free energy density in the isotropic phase and the normalized order parameter, $0 \leq \psi \leq 1$. For the disordered phase, $\psi = 0$ and $\psi = 1$ for a fully ordered phase. The constants, a, b and c depend on the thermodynamic variables of state. From the condition that equilibrium order parameter minimizes the free energy, $(dF/d\psi) = 0$ gives

$$a(T - T^*)\psi + b\psi^2 + c\psi^3 = 0 \quad (8.7)$$

The general solution of order parameter in the ordered phase is

$$\psi = \frac{-b \pm \sqrt{b^2 - 4ac(T - T^*)}}{2c} \quad (8.8)$$

At the phase transition temperature both the phases must have same value of free energy density requires that

$$F(T, \psi) - F_0 = \frac{1}{2}a(T - T^*)\psi^2 + \frac{1}{3}b\psi^3 + \frac{1}{4}c\psi^4 = 0 \quad (8.9)$$

giving

$$\psi = \frac{-2b \pm \sqrt{4b^2 - 18ac(T - T^*)}}{3c} \quad (8.10)$$

From Eqs. 8.8 and 8.10 the value of the order parameter at transition temperature (T_c) can be obtained as

$$\psi = -2b/3c. \quad (8.11)$$

Thus, there is a discontinuity in the order parameter at transition temperature at

$$T_c - T^* = (2b^2/9ac) \quad (8.12)$$

Neglecting the higher order terms of ψ in Eq. 8.7 we get

$$a(T - T^*)\psi + b\psi^2 = 0 \quad (8.13)$$

The value of a/b was obtained from the linear fitting of Eq. 8.13, which was further used in Eq. 8.12 to get b/c.

$$\Delta f = \frac{1}{2}(a/c)(T - T^*)\psi^2 + \frac{1}{3}(b/c)\psi^3 + \frac{1}{4}\psi^4 \quad (8.14)$$

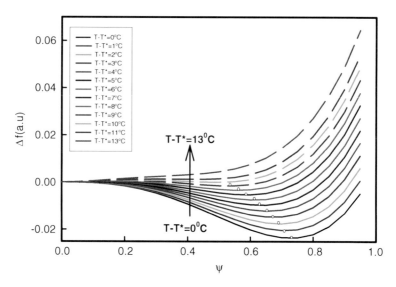

Fig. 8.6 Plot of free-energy as function of order parameter for various (T − T*) values. The *circles* on the *curves* depict the location of minima. For (T − T*) > 10 °C, the system has no minima in free-energy. In contrast, for (T − T*) < 10 °C, the minima shifts to higher order parameter values and gets deeper. Open *circles* represents ψ_0, which is the order parameter corresponding to minimum free-energy

where, $\Delta f = \frac{F(T,\psi) - F_0}{c}$

This enabled us to construct the free-energy plot shown in Fig. 8.6. Consequently, it was possible to plot ψ_0 as function of $(T_c - T)$ which is shown in Fig. 8.7. Least-squares fitting of data yields

$$\psi_o \sim (T_c - T)^{-\alpha}; \quad \alpha = 0.40 \pm 0.05 \tag{8.15}$$

Realize that T_c predicted by Landau theory is same as determined from experiments. Figure 8.2 also reveals the variation of material anisotropy as function of temperature $(T_c - T)$. Moreover, fitting of this data to Eq. 8.3 gives $\beta = 0.40 \pm 0.05$ implying $\alpha \approx \beta$. Thus, it is imperative to establish physical correlation between Landau formalism and the anisotropy data. The solution containing the two clays can be described as an ensemble of interconnected domains with each domain containing a finite number of Laponite-MMT clusters. At room temperature, these domains are completely hydrated and, hence are largely symmetric in structure. Thus, the light scattered from these samples have a smaller component of anisotropy. As the temperature rises, hydration solvent loss causes these domains to gain structural asymmetry which imparts optical anisotropic properties to these moieties. Maximum anisotropy would be observed at the transition temperature T_c. Depolarization ratio is directly associated with particle asymmetry. In this picture, the role of order parameter is adequately played by the anisotropy of the domains (observed as D_p) which explains the observed similarity between temperature dependent behaviour of order parameter with that of anisotropy, see Fig. 8.7.

8.3 Application of Landau Theory

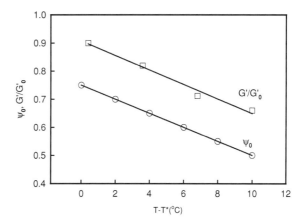

Fig. 8.7 Variation of order parameter and relative storage modulus as function of (T − T*). The storage modulus values are taken from Fig. 8.1 and $G_0'^*$ is G' at T*. The *solid lines* are least squares fitting to the data points that yield universal slope = 0.041. ψ_0 is the order parameter corresponding to minimum free-energy (see Fig. 8.6)

The free-energy is plotted as function of order parameter for various temperatures in Fig. 8.6. For $\psi = 0$, the free-energy curve shows a minima for all T corresponding to a stable reference state. For $T - T^* = 10\,°C$, the free-energy just starts to develop a second minima, that clearly defines the transition temperature, T_c (and $T^* = 50\,°C$). As expected for $T > T_c$ there is no minima.

Thus, the reference temperature T^* lies in the neighborhood of T_c. As one moves away from T_c $(T - T^* < 10\,°C)$ the second free-energy minima shifts towards higher order parameter values and gets deeper until one reaches $T - T^* = 0$. This is clearly seen from Fig. 8.6. Let us look at the cooling cycle data shown in Fig. 8.1. As has been discussed earlier, the dispersion left at T_c spontaneously moves along this branch which implies that there are several physical states with favorable free-energy minima below T_c which is in agreement with the free-energy diagram.

In order to make it amply clear, we plot the values of order parameter corresponding to the various free-energy minima curves and $G'/G_0'^*$ (from Fig. 8.7 data) ($G_0'^* = G'$ at T^*), as function of $(T - T^*)$ that yields an universal slope = 0.041 (Fig. 8.7). Thus, it was concluded that D_p unambiguously and adequately defined the order parameter in our problem and the Landau theory sufficiently described the phase transition observed in these experiments. In Fig. 8.1, both Laponite and MMT dispersions are seen to exhibit dehydration transition like in what is seen in the mixed system. We conducted depolarized light scattering studies on these samples. The Laponite dispersion did not produce reproducible I_{VH} data while for MMT dispersions the variation in I_{VH} as function of temperature was limited to ≈15 %. Thus, it was not possible to apply Landau theory to these dispersions.

8.4 Conclusion

We have undertaken a detailed study on mixed discotic arrested system, comprising of Laponite and MMT, in order to understand the associated dispersion phase behaviour during thermal dehydration. The dispersions formed with equal concentration of both clays had high network rigidity (or yield strength) and was homogeneous in nature at room temperature. This dispersion showed thermally activated irreversible conformational phase transition from mostly isotropic (disordered) to strongly anisotropic (ordered) phase at temperature ≈60 °C. This phase transition could be modeled through Landau formalism wherein it is conceptualized that the comprised hydrated domains of clay clusters that undergo dehydration at higher temperature. We have explicitly shown that the order parameter and the depolarization ratio (domain anisotropy) scale universally with same exponent. Interestingly, the storage modulus followed similar behaviour which may or may not be attributed to mere coincidence. This phase transition was facilitated by the reorganization of the interconnected colloidal domains. In addition, plausibly the glass system is associated with desiccating properties at higher temperature.

The ordered phase was associated with storage modulus values that was several orders of magnitude larger than the preceding phase. In addition, this modulus increased in the cooling cycle implying that the arrested phase was gaining mechanical rigidity and thermal stability substantially in a cooperative manner. This was adequately modeled following Landau theory using experimental depolarization ratio as appropriate order parameter. Moreover, the results on hand establish that a discotic glass has many thermodynamically equilibrium phase states though some of these need to be activated. A pertinent question arises here: is the ordered state like in the nematic phase seen in liquid crystals? There is no obvious answer to this question. No optical birefringence studies were carried out to establish this. Recently, [4] used molecular dynamics to study structural arrest in colloidal suspensions. This study concludes that colloidal arrest is significantly dependent on the formation of long living clusters that span the system which constitutes a new concept. Thus, the field of structural arrest in colloidal systems will continue to attract many studies in the future since these systems are poorly understood.

References

1. P. Bala, B.K. Samantaray, S.K. Srivastava, Dehydration transformation in ca-montmorillonite. Bull. Mater. Sci. **23**, 61–67 (2000)
2. B.J. Berne, R. Pecora, *Dynamic Light Scattering with Applications to Chemistry, Biology and Physics* (Wiley-Interscience, New York, 1976)
3. P.M. Chaikin, T.C. Lubensky, *Principles of Condensed Matter Physics* (Cambridge University Press, Cambridge, 1998)
4. A. de Candia, E. Del Gado, A. Fierro, N. Sator, M. Tarzia, A. Coniglio, Columnar and lamellar phases in attractive colloidal systems. Phys. Rev. E **74**, 010403 (2006)
5. M. Doi, A. Onuki, Dynamic coupling between stress and composition in polymer solutions and blends. J. Phys. II (France) **2**, 1631 (1992)
6. A. Onuki, T. Taniguchi, Viscoelastic effects in early stage phase separation in polymeric systems. J. Chem. Phys. **106**, 5761 (1997)
7. G. Ovarlez, P. Coussot, Physical age of soft-jammed systems. Phys. Rev. E **76**, 011406 (2007)

References

8. B. Ruzicka, L. Zulian, G. Ruocco, More on the phase diagram of laponite. Langmuir **22**, 1106–1111 (2006)
9. B. Ruzicka, L. Zulian, R. Angelini, M. Sztucki, A. Moussaid, G. Ruocco, Arrested state of clay-water suspensions: gel or glass? Phys. Rev. E **77**(2), 020402 (2008)
10. B. Ruzicka, L. Zulian, E. Zaccarelli, R. Angelini, M. Sztucki, A. Moussaid, G. Ruocco, Competing interactions in arrested states of colloidal clays. Phys. Rev. Lett. **104**, 085701 (2010)
11. H. Tanaka, J. Meunier, D. Bonn, Nonergodic states of charged colloidal suspensions: repulsive and attractive glasses and gels. Phys. Rev. E **69**, 031404 (2004)

Chapter 9
Aggregation and Scaling Behavior of Nanoclays in Alcohol Solutions

Abstract This chapter deals with hydrophobic interaction mediated aggregation of clay particles in monohydric alcohol solutions. Nanoclays of different aspect ratio, [Laponite (aspect ratio = 30) (L) and Na-Montmorillonite (aspect ratio = 250) (MMT)] were used to pursue their physical properties in hydrophobic alcohol solutions.

9.1 Introduction

Dispersion stability and phase behavior of clays in aqueous medium have been studied extensively in the literature [23–25]. Very little is known about the physical properties of clay dispersed in non-aqueous solvents. These solvents are characterized by their reduced polarity and finite hydrophobicity. Solvent hydrophobicity plays important role in diverse phenomena ranging from the creation of emulsion to the assembly of biological molecules.

The hydrophobic effect is presumed to be an important driving force in biological and nanoscale self-assembly [2, 12, 28]. Self-organization processes in nature have become increasingly important owing to their relevance in the design and understanding of functional units on a supramolecular level [1, 6–8]. Modeling the hydrophobic effect remains a challenge. However, it is relatively easier to study model hydrophobic systems from an experimental perspective. A common method of investigation of systems where the role of solvent hydrophobicity can determine the nature and extent of interactions, such as in denaturation of nucleic acids and proteins, or surfactant micellization, is to alter the water structure by adding small quantities of monohydric alcohols of different chain lengths [9, 18]. Such alcohol solutions have been studied extensively [4, 13, 30] and their physicochemical behavior in the water-rich region has been probed [14]. Dilute aqueous solutions of monohydric alcohols can be classified as 'hydrophobic systems' even though alcohol molecules are hetero functional (these have both solvophobic and solphophilic groups). Water is a highly

Table 9.1 Comparison of physical properties and scaling exponents for the two clays (Laponite L and MMT) in different alcohol solutions

Clay	ρ (g/cc)	Aspect ratio	D_{eff} (nm)	ζ (mV)	Alcohol	Scaling exponents			
						a	b	c	d
L	2.53	30	30	−40	MOH	0.62	0.73	0.63	
					EOH	0.56	0.67	0.60	1.15
					POH	0.57	0.69	0.57	
MMT	2.86	250	240	−32	MOH	0.63	0.74	0.61	
					EOH	0.58	0.65	0.61	1.15
					POH	0.56	0.670	0.58	

ρ = density of the clay, D_{eff} = effective diameter of the particle, ζ = zeta potential

structured liquid with a local tetrahedral configuration. Mixing of water with simple alcohols result in cross intermolecular associations and deviations from ideal thermodynamic mixing behavior have been observed [5, 19].

Solute-solvent interaction between clay particles and organic solvents may lead to observation of phase states with interesting properties. Dispersion of clay in organic solvents is limited by the poor solubility of these charged colloidal particles. However, aqueous solutions of primary alcohols, like; methanol, ethanol and 1-propanol, provide adequate solubility to clay particles. It was imperative to study the solution state behavior of clay dispersions in these solvents, as this could lead to the possible generation of a new class of soft materials, "clay organogels". In a recent study, thin-shell vesicles were shown to form when clay armored bubbles were exposed to certain water-miscible organic liquids [17, 26]. The interfacial tension of these mixtures is a physical property of great importance because it governs processes in mass transfer such as distillation, extraction, or absorption. Therefore, it is important to study the interaction between the clay particles and the organic liquids like alcohols. Since, clay particles are insoluble in alcohols, and strongly soluble in water, the solvation dynamics of clay in such solutions becomes quite complicated which is poorly understood.

In an earlier work, we had discussed the behavior of macroscopic properties of Laponite dispersed in monohydric alcohols [20]. A pertinent question arises: is there a unified dependence of physical properties of clay dispersions on solvent polarity? We address this issue in this chapter. The aspect ratio plays an important role in the structural arrangement of the nanoparticles in a given solvent. The crystal structure and composition of the clays, used in this study: Laponite (aspect ratio = 30) and Montmorillonite (aspect ratio = 250), is available elsewhere [22, 27]. A brief summary of the physical properties of these clays is provided in Table 9.1.

9.2 Sample Preparation

Very dilute suspensions of the clays were used in this study to avoid initial clustering. Laponite and MMT (each 0.1 % w/v) dispersion was prepared by dissolving Laponite and MMT separately in deionized water at room temperature and the dispersions were sonicated for 10 min. The dispersions were filtered through 0.45 μm Millipore filter. Hydrophobicity gain and polarity loss was introduced into the solvent by adding primary alcohols (methanol, ethanol and 1-propanol) in different proportions to water.

Hydrophobic aggregation of clays of different aspect ratios in three different alcohol (methanol, ethanol, and 1-propanol) solutions was investigated through zeta potential, hydrodynamic size, viscosity and surface tension measurements.

9.3 Results and Discussion

Water is a poor diluent for organic molecules. This can be improved by addition of co-solvents to water; thereby enhancing the solubility of these organic molecules. Among such additives primary alcohols stand out in all aspects as unique co-solvents. In this study, three hydrogen bond donating solvents (primary alcohols) were chosen as model dispersion media. These were mixed with water to generate alcohol solutions of desired alcohol concentration of mole fraction χ_{OH}. When alcohol concentration is raised, the polarity of the water-alcohol binary mixture is reduced and its hydrophobicity is increased. Thus, by changing χ_{OH}, it is possible to control the solvent polarity and hydrophobicity. The solvent polarity is measured by the π^* solvatochromic scale [11]. The π^* scale was originally constructed in order to represent the nonspecific polarity/polarizability of solvents, exclusive of the influence of solvent-solute hydrogen-bonding and other specific interactions. However, it was realized that π^* scale inadequately defined hydrogen-bonding solvents. A more detailed theoretical model replaced π^* polarity by a dielectric measure of solvent polarity, called reaction field factor, defined as [11]

$$f(\varepsilon_0, n) = \left(\frac{\varepsilon_0 - 1}{\varepsilon_0 + 2}\right) - \left(\frac{n^2 - 1}{n^2 + 2}\right) \tag{9.1}$$

where ε_0 and n are the static dielectric constant and refractive index of the solvent respectively. This reaction field factor comes directly from the dielectric continuum theory [3, 15, 16]. This theory mainly assumes that a point dipole solute interacts with the solvent by virtue of the change in solute dipole moment.

Relative polarity of the alcohol solution, $f_{soln}(\varepsilon_0, n)$, with respect to that of water, $f_w(\varepsilon_0, n)$, can be defined as

$$\delta f = f_w(\varepsilon_0, n) - f_{soln}(\varepsilon_0, n) \tag{9.2}$$

The parameter δf will be used henceforth to describe solvent polarity and hydrophobicity. It was determined by using the dielectric constant and refractive index of different alcohol solution from literature [31]. The zeta potential for the two clays (laponite and MMT) was determined as function of hydrophobicity and polarity of the solvent which was changed systematically by varying concentration of alcohol in water. Let ζ_{OH} and ζ_w be the zeta potential of the two clays in alcohol solution and water respectively.

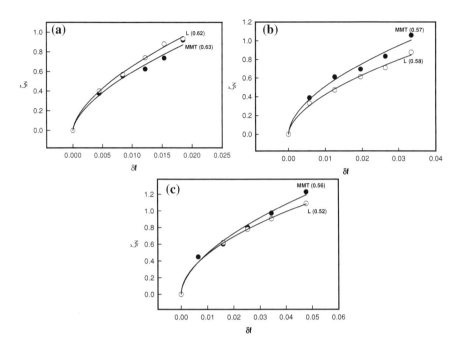

Fig. 9.1 Variation of normalized zeta potential, ζ_N for two clays (laponite L and MMT) shown as function of solvent polarity for **a** methanol, **b** ethanol and **c** 1-propanol solutions. *Solid line* represents fitting of data to the expression given in Eq. 9.3. The power-law exponents obtained is shown on the *curves*

The normalized zeta potential (ζ_N) data for the two clays is shown in Fig. 9.1 and the data could be fitted to the power-law relation

$$\frac{\zeta_{OH} - \zeta_W}{\zeta_W} = \zeta_N \sim \delta f^a \qquad (9.3)$$

Values of the exponent obtained from fitting were, a = 0.62 ± 0.03, 0.56 ± 0.02, 0.57 ± 0.03, same for both clays in methanol, ethanol and 1-propanol solutions respectively.

The medium dielectric constant reduces with increasing alcohol concentration (or decreasing polarity of the solvent) which favors attractive electrostatic interaction between the face of one and rim of another disc. Figure 9.2 clearly shows the increase in effective hydrodynamic size of colloidal cluster of both clays with change in polarity of the solvent. The normalized growth in cluster size could be given by a power-law as

$$\frac{R_{OH} - R_W}{R_W} = R_N \sim \delta f^b \qquad (9.4)$$

where R_{OH} and R_W are the effective hydrodynamic radius of the clay clusters in alcohol solution and water respectively. Values of the exponent obtained from

9.3 Results and Discussion

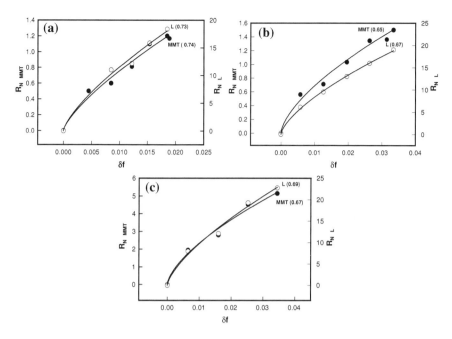

Fig. 9.2 Variation of normalized size for two clays (laponite and MMT) as function of polarity of **a** methanol, **b** ethanol and **c** 1-propanol solutions. *Solid line* represents fitting of data to the expression given in Eq. 9.4. The power-law exponents obtained is shown on the *curves*

fitting were $b = 0.73 \pm 0.03$, 0.65 ± 0.03, 0.67 ± 0.03, same for both clays in methanol, ethanol and 1-propanol solutions respectively.

The viscosity data for the dispersions is shown in Fig. 9.3. The normalized viscosity (η_N) for the two clays in dispersions showed a power-law dependence on solvent polarity (δf) as

$$\frac{\eta_{OH} - \eta_W}{\eta_W} = \eta_N \sim \delta f^c \qquad (9.5)$$

where η_{OH} and η_W are the viscosity of the alcohol solution containing clay particles and water respectively. The raw data of viscosity is shown in Fig. A.1 in the Appendix. Data fitting yielded values for $c = 0.61 \pm 0.02$, 0.61 ± 0.02, 0.57 ± 0.02 same for both the clays in methanol, ethanol and 1-propanol solutions respectively.

The normalized surface tension (γ_N) for two clays in the ethanol solution is shown in Fig. 9.4. This data was least-squares fitted to the following power-law expression

$$\frac{\gamma_{OH} - \gamma_W}{\gamma_W} = \gamma_N \sim \delta f^d \qquad (9.6)$$

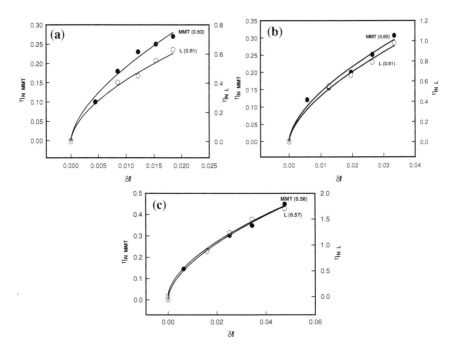

Fig. 9.3 Variation of normalized viscosity for clay dispersions as function of polarity of **a** methanol, **b** ethanol and **c** 1-propanol solutions. *Solid lines* represents fitting of data to the expression given in Eq. 9.5. The power-law exponents obtained is shown on the *curves*

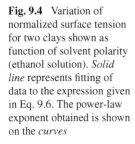

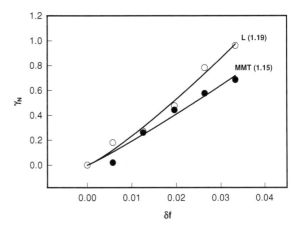

Fig. 9.4 Variation of normalized surface tension for two clays shown as function of solvent polarity (ethanol solution). *Solid line* represents fitting of data to the expression given in Eq. 9.6. The power-law exponent obtained is shown on the *curves*

where γ_{OH} and γ_W shows the surface tension of the clay dispersion in solution and water respectively. Value of exponent found was $d = 1.1 \pm 0.1$ same for both the clays.

Interestingly, the power law exponents a, b, c and d were same for both clays irrespective of the fact that the two clays were structurally different. In other

9.3 Results and Discussion

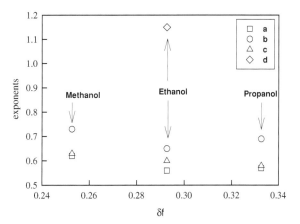

Fig. 9.5 Variation of various scaling exponents obtained from different alcohol solutions as a function of solvent polarity

words, discotic colloids with different aspect ratio have exhibited unified dependence on polarity and hydrophobicity of the dispersion medium. If Z represents the physical property, like, the zeta potential, apparent hydrodynamic radius, viscosity and surface tension, of the clay dispersion in different alcohol solutions then it was observed that Z scaled with the solvent polarity (δf) as $Z \sim \delta f^\alpha$ where α is the exponent which was different for different physical properties, but same for both clays. Table 9.1 summarizes the various properties of the clays and the observed exponents in different alcohol solutions (Fig. 9.5).

It is well known that solvation of reactants significantly affects rates of reactions. In binary solvents this process is altered considerably due to the occurrence of preferential solvation. The results, on hand, imply a manifestation of this phenomenon in alcohol solutions though there were no chemical reactions involved. However, the selective physical hydration of the clay aggregates appears to be self-organized in a manner that attributed unified physical properties to these dispersions.

The aforesaid results clearly implied unified behavior of the macroscopic physical properties of the two clays as a function of solvent polarity and/or hydrophobicity. This behavior depends on the various interactions prevailing in the system which includes solvent-solvent, solute-solvent and solute-solute interactions. The solvent-solvent interaction in water-alcohol binary solvent is mainly dominated by the hydrogen bonding as alcohols are hydrogen bond donors and water is an excellent hydrogen bond acceptor. It has been reported that [10, 21] water alcohol interact via hydrogen bond to form a clathrate structures and in their formation first hydration shell occur when $0.05 < \chi_{OH} < 0.1$ for both methanol and ethanol. As $\chi_{OH} \geq 0.3$ hydrophobic hydration becomes negligible which is indicated from minimum self diffusion coefficient of water. Hence water structure is completely lost and water molecules mix freely with alcohol. At high χ_{OH} values (or lower polarity of the solvent), solution dynamics of water and alcohol becomes completely independent. The binary solvent rich in intermolecular hydrogen bond act as a marginal solvent for the clay particles. For higher polarity (or $\chi_{OH} < 0.3$) more of water is available to the clay particles for their dispersion in other words solute-solute interaction dominates over the solute-solvent part. As the hydrophobicity increases (by increasing the alcohol concentration) solute-solvent interaction becomes less and solute-solute interaction increases which results in the aggregation of clay particles (this is clear from Fig. 9.2). The similarity in the hydrophobic aggregation of the two clays of different aspect ratio can be explained by the similar solute-solvent interaction for the two types of clay. Similar observation has been made by Tuulmets et al. [29]. They found significant similarity between the solute-solvent interactions in wide range of co-solvent concentration.

9.4 Conclusion

In summary, different physical properties including zeta potential, apparent hydrodynamic radius, viscosity and surface tension of clay dispersion of different aspect ratios have been systematically studied as function of solvent polarity and hydrophobicity. The physical properties (Z) of both clays were found to show the power law behavior with the solvent polarity (hydrophobicity) as $Z \sim \delta f^\alpha$ where α was exponent which was found to be different for different physical properties, but same for both clays. The unified scaling behavior observed implies that solute-solvent interaction between the clay and alcohol solutions is governed by a dispersion dynamics that is invariant of the particle aspect ratio. Even though the anisotropic particles of different aspect ratio behave differently their aggregate formation induced by the hydrophobic interactions follow the same path. It may be due to the formation of spherical clusters which are symmetric in nature.

References

1. H. Acharya, S. Vembanur, S.N. Jamadagni, S. Garde, Mapping hydrophobicity at the nanoscale: applications to heterogeneous surfaces and proteins. Faraday Discuss. **146**, 353–365 (2010)
2. B. Alberts, A. Johnson, J. Lewis, M. Raff, K. Roberts, P. Walter, *Molecular Biology of the Cell*, 5th edn. (Garland Science, New York, 2007)
3. A.T. Amos, B.L. Burrows, Solvent-shift effects on electronic spectra and excited-state dipole moments and polarizabilities. Adv. Quantum Chem. **7**, 289 (1973)
4. M.D. Angelo, G. Onori, A. Santucci, Self-association of monohydric alcohols in water: compressibility and infrared absorption measurements. J. Chem. Phys. **100**, 3107–3113 (1994)
5. F. Biscay, A. Ghoufi, P. Malfreyt, Surface tension of water–alcohol mixtures from Monte Carlo simulations. J. Chem. Phys. **134**, 044709 (2001)
6. D. Chandler, Interfaces and the driving force of hydrophobic assembly. Nature **437**, 640–647 (2005)
7. N.V. Chennamsetty, V. Voynov, B. Helk Kayser, B.L. Trout, Design of therapeutic proteins with enhanced stability. Proc. Natl. Acad. Sci. U.S.A. **106**, 11937 (2009)
8. N. Chennamsetty, V. Voynov, V. Kayser, B. Helk, B.L. Trout, Prediction of aggregation prone regions of therapeutic proteins. J. Phys. Chem. B. **114**, 6614 (2010)
9. S. Cinelli, G. Onori, A. Santucci, Effect of 1-alcohols on micelle formation and protein folding. Colloids Surf. A. Physicochem. Eng. Aspects. **160**, 3–8 (1999)

References

10. S. Dixit, J. Crain, W.C.K. Poon, J.L. Finney, A.K. Soper, Molecular segregation observed in a concentrated alcohol–water solution. Nature **416**, 829 (2002)
11. M.L. Horng, J.A. Gardecki, A. Papazyan, M. Maroncelli, Subpicosecond measurements of polar solvation dynamics: coumarin 153 revisited. J. Phys. Chem. **99**, 17311 (1995)
12. W. Kauzmann, Some factors in the interpretation of protein denaturation. Adv. Protein Chem. **14**, 1–63 (1959)
13. G.C. Kresheck, Surfactants, in *Water: A Comprehensive Treatise*, vol. 4, ed. by F. Franks (Plenum Press, New York and London, 1975), p. 95
14. A. Laaksonen, P.G. Kusalik, I.M. Svishchev, Three-dimensional structure in water-Methanol mixtures. J. Phys. Chem. A **101**, 5910 (1997)
15. N. Mataga, T. Kubota, *Macromolecular interaction and electronic spectra* (Mercel Dekker, New York, 1970), p. 371
16. E.G. McRae, Theory of solvent effects on molecular electronic spectra frequency shifts. J. Phys. Chem. **61**, 562 (1957)
17. G. Odriozola, A. Schmitt, J. Callejas-Fernández, R. Hidalgo-Álvarez, Aggregation kinetics of latex microspheres in alcohol–water media. J. Colloid Interface Sci. **310**, 471 (2007)
18. G. Onori, A. Santucci, Dynamical and structural properties of water/alcohol mixtures. J. Mol. Liq. **69**, 161 (1996)
19. S.A. Patel, C.L. Brooks III, Structure, thermodynamics, and liquid-vapor equilibrium of ethanol from molecular-dynamics simulations using nonadditive interactions. J. Chem. Phys. **123**, 164502 (2005)
20. N. Pawar, H.B. Bohidar, Hydrophobic hydration mediated universal self-association of colloidal nanoclay particles. Colloid Surf A: Physicochem Eng. Aspects. **333**, 120–125 (2009)
21. W.S. Price, H. Ide, Y. Arata, Solution dynamics in aqueous monohydric alcohol systems. J. Phys. Chem. A **107**, 4784 (2003)
22. R.K. Pujala, N. Pawar, H.B. Bohidar, Universal sol state behavior and gelation kinetics in mixed clay dispersions. Langmuir **27**, 5193 (2011)
23. R.K. Pujala, H.B. Bohidar, Slow dynamics, hydration and heterogeneity in laponite dispersions. Soft Matter **9**, 2003–2010 (2013)
24. B.L. Ruzicka, L. Zulian, G. Ruocco, More on the phase diagram of laponite. Langmuir **22**, 1106–1111 (2006)
25. B. Ruzicka, E. Zaccarelli, L. Zulian, R. Angelini, M. Sztucki, A. Moussaïd, T. Narayanan, F. Sciortino, Observation of empty liquids and equilibrium gels in a colloidal clay. Nat. Mater. **10**, 56–60 (2011)
26. A.B. Subramaniam, J. Wan, A. Gopinath, H.A. Stone, Semi-permeable vesicles composed of natural clay. Soft Matter **7**, 2600 (2011)
27. S.L. Swartzen-Allen, E. Matijevic, Surface and colloid chemistry of clays. Chem. Rev. **74**, 385 (1974)
28. C. Tanford, *The Hydrophobic Effect: Formation of Micelles and Biological Membranes* (Wiley, New York, 1973)
29. A. Tuulmets, J. Jarv, T. Tenno, S. Salmar, Significance of hydrophobic interactions in water–organic binary solvents. J. Mol. Liq. **148**, 94 (2009)
30. P. Varilly, A.J. Patel, D. Chandler, An improved coarse-grained model of solvation and the hydrophobic effect. J. Chem. Phys. **134**, 074109 (2010)
31. CH. Wohlfarth, Dielectric constant of the mixture (1) water; (2) methanol, in *Springer Materials—The Landolt-Börnstein Database*, ed. by M.D Lechner (Springer, Berlin, 2008) (http://www.springermaterials.com). doi: 10.1007/978-3-540-75506-7_321

Chapter 10
Summary

Abstract In this concluding chapter, I summarize and review the main findings of this thesis and describe a series of opportunities for interesting future work on similar systems.

10.1 Summary of the Main Results

This thesis has explored the phase behavior of colloidal nano clays of different aspect ratio in the individual and their mixed states. The nanodiscs having anisotropic interactions produced variety of phase states depending on the concentration and waiting time. Colloidal gels have formed via different routes such as phase separation and equilibrium. One of the significant findings of the thesis is that aging and temperature drive the system towards its minimum free energy or equilibrium configuration. Wide range of experimental techniques and a few theoretical treatments were employed to understand the formation of different distinguished phases in nanodisc dispersions. The important findings of this thesis are reviewed as follows:

- **Phase diagram of aging Laponite in aqueous medium**. Laponite dispersions in water at room temperature in the concentration range 0.1–3.5 % (w/v) were examined by an array of experimental tools over a period of one year. The temporal growth of the self-assembled colloidal structures was probed by light scattering studies which revealed the existence of aggregating structures in both sol and gel states as opposed to the glass phase where the presence of an amorphous arrested phase was indicated. Both in the sol and gel regime $[c < 1.8~\%~(w/v)]$, the intensity of light scattered $I(q, c)$ scaled with concentration c as, $I(q, c) \sim c^\alpha$ with $\alpha = 0.95$ at $t_w = 0$, and 0.63 after $t_w = 6$ months implying that this temporal growth resulted from the formation of colloidal gel whereas in the glass phase $[c \geq 2~\%~(w/v)]$ scattered intensity from samples remained constant ($\alpha = 0$). The hydration of these anisotropic charged platelets was studied using Raman spectroscopy in a systematic and structured manner over the same period which unambiguously established the presence of three phase states of the dispersion, with each of these aging in its characteristic way. This data was supplemented by the information obtained from rheological mapping of the samples. A time-dependent 3D-phase diagram has been proposed for the salt-free Laponite dispersion in water which clearly shows distinct clear *sol*, *gel* and *glass* states.

- **Phase diagram of aging Montmorillonite dispersions**. We have investigated the spontaneous evolution of various self-assembled phases from a homogeneous aqueous dispersion of high-aspect ratio Sodium Montmorillonite (*Na* Cloisite) nanoclay platelets based on the observations made over a period of 3.5 years. We have established for the first time ever the t_w-c phase diagram for this system in salt-free suspensions under normal pH conditions using rheology and have detected that these suspensions do undergo nontrivial phase evolution and aging dynamics. Distinctive *phase separation, equilibrium fluid* and *equilibrium gels* in the $t_w - c$ phase space were discovered. Cole-Cole plots derived from rheology measurements suggested the presence of inter connected network-like structures for $c > c_g$, c_g being the gelation concentration. During the initial time all dispersions formed stable sols, and with aging network-like structures were found to form via two routes: one for $c < c_g$, by *phase separation* and another for $c > c_g$, through *equilibrium gelation*. The yield stress of the Laponite and MMT grows as cube of the concentration of clay concentration which indicates the universal behavior in anisotropic colloids with complex interactions.

- **Orientational ordering of Laponite particles induced by water-air interface**. We studied the kinetics of interface induced generation, propagation of arrested phase caused anisotropy, packing and relaxation dynamics in Laponite. Ordering effects and changes in the slow dynamics were detected by a specially designed depolarized light scattering setup that allowed measurement at different distances from the interface. The geometry that we have used in our experiments conclusively mapped the dynamics of anisotropy growth occurring at the interface and propagating into the bulk with aging. We believe that the water-air interface is responsible for causing this ordering. The density fluctuations at the interface of water-air are much stronger compared to the bulk, which may be responsible for the observed occurrence of anisotropy. The parameters that enhanced the ordering were the aging time and temperature. Higher the temperature higher was the amount of depolarization, and the particles encountered arrested state early. Interestingly, we noticed the presence of heterogeneous relaxations in these dispersions that varied with depth which was not seen hitherto. When the interface was covered with oil or water immiscible liquids, the surface fluctuations at the interface vanished and the abovementioned dynamics of aging was absent. Covering the surface also delayed the aging process. We infer that pH and salt dependence have to be studied in this system to gain further insight into the ordering phenomenon.

10.1 Summary of the Main Results

- **Sol behavior gelation kinetics in mixed nanoclay dispersions of Laponite and Montmorillonite**. We undertook a detailed study on discotic sols and gels, comprising of Laponite and MMT, in order to understand the associated sol and gel phase behaviour during pre and post gelation scenario. The sol state exhibited zeta potential and gelation concentration values that could be expressed as linear combination of the same of their constituents. Clay dispersions prepared with varying mixing ratios showed enhanced mechanical and thermal properties. The mixed discotic colloidal systems were found to be having gel-like features: well defined gelation concentration and finite rigidity modulus. In addition, the experimental data, in hand, mostly supported percolation type gelation mechanism. At the same time, it was possible to note several universal features both in sol and gel states. Thus, the mixed clay sol transforms to a colloidal gel alike in the gelation of polymers. The dispersions formed with equal concentration of both clays had high network rigidity (or yield strength) and was homogeneous at room temperature. The hardening transitions were studied systematically for the first time in the dispersions of Laponite, MMT and the mixtures of the same.

- **Aging dynamics in mixed nanoclay dispersions**. We systematically probed the slow dynamics in aging Laponite-MMT glass system at room temperature. Interestingly, the two key characteristic parameters defining the aging of the system like, the fast and slow mode relaxation time of the dynamic structure factor, and the concentration dependence of ergodicity breaking time, exactly followed the trend observed in Laponite glass. It was observed that the ergodicity breaking time was an exponentially decaying function of clay concentration alike what is seen in Laponite glass. Thus, it could be unambiguously concluded that the aging pathway observed in Laponite-MMT glass followed qualitatively same footprints of Laponite glass. Consequently, all the inter-platelet interactions was primarily dominated and governed by Laponite-Laponite electrostatic forces in the mean field provided by the weakly charged MMT platelets. This attributed a matrix-like platform provided by MMT on which the Laponite particles strongly interacted. Further experiments carried out identified the said phases as glass-like. In order to confirm that the arrested phase was indeed a glass phase, dilution tests were carried out on these samples which established the presence of glass phase in our system.

- **Thermally induced irreversible ordering in mixed nanoclay dispersions**. We have undertaken a detailed study on mixed discotic arrested system, comprising of Laponite and MMT, in order to understand the associated dispersion phase behaviour during thermal dehydration. The dispersions formed with equal concentration of both clays had high network rigidity (or yield strength) and was homogeneous in nature at room temperature. This dispersion showed thermally activated irreversible conformational phase transition from mostly isotropic (disordered) to strongly anisotropic (ordered) phase at temperature $\approx 60\ °C$. This phase transition could be modeled through Landau formalism wherein it was conceptualized that the aggregates comprised hydrated domains of clay clusters that undergo dehydration at higher temperature. We have explicitly shown that the order parameter and the depolarization ratio (domain anisotropy) scale universally with same exponent. Interestingly, the storage modulus follows similar behaviour which may or may not be attributed to mere coincidence. This phase transition was facilitated by the reorganization of the interconnected colloidal domains. The ordered phase was associated with storage modulus value that was several orders of magnitude larger than the preceding phase. In addition, this modulus increased in the cooling cycle implying that the arrested phase was gaining mechanical rigidity and thermal stability substantially in a cooperative manner. These ordered state may be alike the nematic phase seen in liquid crystals. One has to carry out optical birefringence studies to establish this.

- **Aggregation and scaling behavior of nanoclays having different aspect ratio in alcohol solutions**. Different physical properties including zeta potential, apparent hydrodynamic radius, viscosity and surface tension of clay dispersion of different aspect ratios have been systematically studied as function of solvent polarity and hydrophobicity. The physical properties (Z) of both clays were found to show the power law behavior with the solvent polarity (hydrophobicity) as $Z \sim \delta f^\alpha$ where α was exponent which was found to be different for different physical properties, but same for both clays. The unified scaling behavior observed implied that solute-solvent interaction between the clay and alcohol solutions was governed by a dispersion dynamics that was invariant of the particle aspect ratio. Even though the anisotropic particles of different aspect ratio behave differently their aggregate formation induced by the hydrophobic interactions follow the same path. It may be due to the formation of spherical clusters which are symmetric in nature. Finally, we concluded with a statement: the physical properties of colloidal dispersions are affected by waiting time (aging), concentration, temperature and hydrophobicity.

10.2 Open Problems

We have identified some of the potential future problems after working in this specific area.

1. **Phase diagram of mixed nano clays of different aspect ratio**. Mixture of particles of different anisotropy may influence the overall phase diagram. For example, we have prepared a sample of equal mixing ratio of Laponite and MMT [0.5 % (w/v)] and to our amazement the so called phase separation did not take place even 4 years after the preparation and it turned out to be viscous. Thus, it will be interesting to study the phase states as function of solid content and waiting time (Fig. 10.1).
2. **Dynamics of nanoclays in hydrophobic environment**. It has been observed that the hydrophobicity increased the yield of the clay dispersions particularly in alcohol solutions. Scaling behavior of yield stress as a function of hydrophobicity was found. It may be the similar case with ionic liquids.

10.2 Open Problems

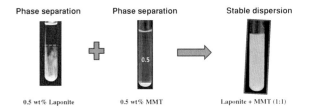

Fig. 10.1 The photographs 0.5 wt% of Laponite and MMT shows the phase separation after 4 years, but the picture at the *right* evolved from the mixed dispersion will not phase separate even after 4 years

3. **Covering the interface surface with water-immiscible solvents**. Through many observations we found that when the interface was insulated with thin layer water-immiscible solvents it yielded very interesting dynamics. The aging dynamics are homogeneous and we can even causes delay the aging dynamics and the ergodicity breaking time. This is an open problem that needs better explanation and further studies to establish it.
4. **Clays as desiccants**. Clays have the ability to absorb the water molecules and reduce the humidity and releases the water at higher temperatures. Thus, clays are widely used as common desiccants. Our studies found that the desiccation temperature for MMT and Laponite are 45 and 40 °C respectively from the rheology measurements. Interestingly, the sample containing the mixture of the two in the 1:1 ratio provided a desiccation temperature 65 °C, which is very good in high temperature regions. Thus it is worthwhile to explore the property of the clay mixtures to make customized materials.
5. **Rheological properties of the mixed nanoclay system**: This study has remained unexplored in this thesis. The mixture of particles exhibit interesting rheological behavior like shear-banding, aging dynamics, relaxation dynamics etc. A feature of shear banding was observed while performing the flow measurements. Thus the work of this thesis gives scope to many future challenging problems.

Appendix

Characterization of Nanoclays

Characterization of Laponite

Dilute solutions of Laponite (0.075 %w/v) were studied by dynamic light scattering and electrophoresis in order to ascertain their physical dimensions and surface potential. Laponite was found to have an effective hydrodynamic radius $\approx 18 \pm 3$ nm. Predominantly negative zeta potential was found to be associated with Laponite particles. The disc (diameter = 30 nm) shaped Laponite particle is known to have a predominantly charged negative surface and a very weakly charged positive rim. Since, electrophoretic measurements estimate only the net charge on the surface of a particle one expects to measure a negative zeta potential for Laponite which was found to be ≈ -40 mV.

Characterization of Na MMT

It was felt necessary to characterize the MMT samples to ascertain their particle size. Among the well-known colloidal models, the one introduced by Hubbard and Douglas [1] seems to be the more realistic approximation applicable to our system. Particles are described as "coin-like" and the diffusion coefficient of this kind of particles is calculated, passing from a hydrodynamic system to a capacitive model, with the formula

$$D_i^0 = \frac{k_B T}{6\pi \eta_0 C_i^0}$$

with D_i^0 being the diffusion coefficient of the object, C_i^0 the equivalent condenser capacity, k_B the Boltzmann constant, T the temperature in Kelvin, and η_0 the solvent viscosity. The equivalent ellipsoid capacity can then be defined with its

Fig. A.1 XRD profiles of Laponite and MMT thin films.

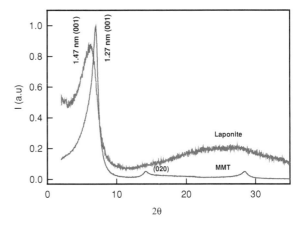

three-dimensional half-lengths a_0, b_0, and c_0. In our "coin-like" geometry, a_0 and b_0 are equal to the radius of the disc surface and c_0 is the half height of the coin $(a_0 = b_0 \neq c_0)$. For $\left(\sqrt{a_0^2 - c_0^2}\right) > 0$, the relation is then

$$C^0 = \frac{\sqrt{a_0^2 - c_0^2}}{\arctan\sqrt{a_0^2 - c_0^2}/c_0}$$

This is used to check the validity of this model and the accuracy of the diffusion coefficients measured by DLS for natural montmorillonite.

The MMT sample used had a concentration 0.01 % (w/v). From dynamic light scattering measurement, $C^0 = 153$ nm (effective hydrodynamic radius) was obtained measured using diffusion coefficient. From literature, one gets thickness of MMT plates, $2c_0 \approx 1.0$ nm. Thus, it was possible to deduce the parameters $a_0 = b_0 \approx 240$ nm (using Eq. 2.14). Thus, the size obtained was consistent with the literature data [2]. The zeta potential of MMT clay was determined to be -30 mV which is not too different from what is reported in Durán et al. [3].

XRD of Clays

Using these XRD data, inter planar spacing was determined for Laponite and MMT thin films. In case of MMT, this spacing was close to 1.27 nm while the Laponite inter planar spacing was 1.47 nm (Fig A.1).

Source of Permanent Charge Cation Exchange Capacity - Montmorilonite

Substitution of Mg^{2+}(&K^+) for Al^{3+}

	Unsubstituted (Pyrophyllite)		Substituted* (Montmorillonite)	
	Ions	Charges + −	Ions	Charges + −
Silica Layer	$6O^{2-}$	12	$6O^{2-}$	12
	$4Si^{4+}$	16	$4Si^{4+}$	16
	$4O^{2-},2OH^-$	10	$4O^{2-},2OH^-$	10
Alumina Layer	$4Al^{3+}$	12	*$2Al^{3+}$, $2Mg^{2+}$ ($2K^+$)	10
	$4O^{2-},2OH^-$	10	$4O^{2-},2OH^-$	10
Silica Layer	$4Si^{4+}$	16	$4Si^{4+}$	16
	$6O^{2-}$	12	$6O^{2-}$	12
		44 44 0		42 44 2(-)

References

1. J.B. Hubbard, J.F. Douglas, Hydrodynamic friction of arbitrarily shaped Brownian particles. Phys. Rev. E **47**, 2983–2986 (1993)
2. A. Cadene, S. Durand-Vidal, P. Turq, J. Brendle, Study of individual Na-montmorillonite particles size, morphology, and apparent charge. J. Colloid Inter. Sci. **285**, 719–730 (2005)
3. J.D.G. Durán, M.M. Ramos-Tejada, F.J. Arroyo, F. González-Caballero, Rheological and electrokinetic properties of sodium montmorillonite suspensions—I. Rheological properties and interparticle energy of interaction. J. Colloid Inter. Sci. **107**, 117 (2000)

Curriculum Vitae

Dr. Ravi Kumar Pujala
Senior Research Fellow
School of Physical Sciences
Jawaharlal Nehru University, New Delhi-110067,
   India
pujalaravikumar@gmail.com

Research Interests

Soft Matter, colloids, polymers, biomolecules, aggregates, gels, glasses, dynamics of anisotropic particles, interface dynamics, nanomaterials and composites, scattering techniques and rheology.

Education

PhD (2013) from School of Physical Sciences, Jawaharlal Nehru University, New Delhi, India

   Thesis Title: *Dispersion Stability, Microstructure and Phase Transition of Anisotropic Nanodiscs*
   Supervisor: Prof. Himadri B. Bohidar

Pre-PhD (2008–2009) School of Physical Sciences, JNU, New Delhi, India; CGPA: 8.4/9.0

Other Competitive Exams Cleared at National level:

   JEST: 92.67 % in Physics (February 2008)
   NET: Awarded CSIR-JRF in Physics (Dec 2007), CSIR-SRF (Ext) (2014)
   IIT JAM: Qualified (2006)

M.Sc. Physics (2008), SPS, JNU, 1st division with CGPA: 6.74/9.0
   Project worked on:

"*Study of Fluorescence from Ag nanoparticles coated with various capping layers*" under the guidance of Prof. Prasenjit Sen, in winter semester 2008.

Summer Project (2007)

"*Synthesis of Molecular electronic switches*" under the guidance of Dr. Pritam Mukhopadhyay.

B.Sc.: Physics, Maths, Chemistry (2006) APRDC, N'Sagar, Acharya Nagarjuna University; 1st Division; 90 %

Intermediate: Physics, Maths, Chemistry (2003) RVVN College, Acharya Nagarjuna University; 1st Division; 92.2 %

SSC: St' Mary's High School. 1st Division; 85 %

Teaching Experience

- Quantum Mechanics I, Monsoon semester 2009
- Quantum Mechanics II, Winter semester 2010
- Electromagnetic Theory, Monsoon semester 2010

Books/Workshop/Oral/Poster Presentation

- **PhD thesis** has been nominated as the best thesis of the year from the department to publish in *The Book Series SPRINGER THESES*.
- *Poster Presentation* on "*Ergodicity breaking and aging dynamics in mixed nanoclay dispersions*" **RK Pujala** and H. B. Bohidar, Julich Soft Matter Days 2012, Nov 13–16, by Institute of complex systems, Julich, Germany.
- DST SERC School and Symposium on "*Rheology of Complex fluids* 2012" by IIT Guwahati, Guwahati, India from January 3–7, 2012.
- *Poster Presentation* on "*Aging and Slow dynamics in Nanoclay dispersions*" **RK Pujala** and H. B. Bohidar, ICONSAT-2012 from January 20–23, 2012, Hyderabad, India.
- *Poster Presentation* on "*Aggregation dynamics of Nano clay particles*" **RK Pujala** and H. B. Bohidar, Indo-US Bilateral Workshop on Nanoparticle Assembly: From Fundamentals to Applications from 12–14 December 2011 organized by IIT Delhi, Delhi, India.
- *Oral and Poster Presentations* on "*Aging transitions in anisotropic colloids*" **RK Pujala** and H. B. Bohidar, Conference and School on Nucleation, Aggregation and Growth, July 26 to August 6, 2010, Bengaluru, India.
- *Oral and Poster Presentation* on "*Phase Behavior of the Nanoclay-Water system*" **RK Pujala** and H. B. Bohidar, National Symposium on Nano science: Theory and Applications, 5–7 November 2009, SES JNU, New Delhi, India
- *Oral presentation* on "*Slow dynamics and Aging in complex fluids*" **RK Pujala**, 7th Dynamics Day, Delhi, 13 December 2011, JNU, New Delhi, India.

- **DSC/TGA Training course** at Perkin Elmer Company from 24th to 26th November, 2010, Mumbai, India.
- **Workshop** on *"Electron Microscopy and its Applications"* by Advance Instrumentation Research Facility, 28–29 February 2012, AIRF, JNU, New Delhi, India.

Laboratory Experience

1. Size, morphology Measurements: By Atomic Force Microscopy (AFM), Scanning Electron Microscope (SEM) and TEM.
2. Surface charge measurement (Zeta Potential): by ZEECOM.
3. Size, diffusion coefficient, radius of gyration, relaxation dynamics of disordered materials composed of nanoparticles: By Dynamic and Static Light Scattering.
4. Hydration of nanoplatelets and structure of water: by Raman Spectroscopy.
5. Basal or interlayer spacing of nanoclays and nanocomposites: by X-ray Diffraction (XRD).
6. Viscoelastic properties, aging dynamics and microstructure: by Rheology.
7. Preparation of metal nanoparticles: by Electro Exploding wire technique.
8. Quantitative determination of absorption of UV light by nanoparticles and biomolecules: by UV spectrophotometer
9. Enthalpic changes in the system: by Differential Scanning Calorimetry (DSC)
10. Interaction and aggregation dynamics: by Colorimeter for Turbid metric measurement.
11. Interaction of fluorescently labeled nanoparticles and their microscopy: by Confocal Microscopy.
12. Diffusion coefficient, size and anisotropy of nanoparticles: by Fluorescence Correlation Spectroscopy (FCS).
13. Chemical synthesis of nanoparticles, chromatographic techniques.

Computer Skill

- Proficient in FORTRAN, MATLAB, MATHEMATICA, Adobe Illustrator etc.
- Working experiences on DOS/Windows/UNIX environment

Awards/Achievements

- CSIR-SRF (2010–2013) (Council of Scientific and Industrial Research-Senior Research Fellowship)
- CSIR-JRF (2008–2010) (Council of Scientific and Industrial Research-Junior Research Fellowship)
- National Eligibility Test (NET) in Physics, conducted jointly by Council of Scientific and Industrial Research (CSIR) and University Grants Commission (UGC), Govt. of India (2007).
- Merit Cum Means scholarship by Jawaharlal Nehru University during MSc
- Dr. Chaitanya Murali scholarship for meritorious students in Post Graduation (2006–2008)
- Best Poster award for presenting the poster entitled "Aging and Slow dynamics in Nanoclay dispersions" in National conference on Supramolecular Chemistry: from Molecules to materials, SPS, JNU, New Delhi, India.

- Young Researcher award and free participation in ICONSAT-2012 from January 20–23, 2012, Hyderabad, India.
- Awarded JNU grant for attending Julich Soft Matter Days 2012, Nov 13–16, by Institute of complex systems, Julich, Germany.
- Awarded Travel grant and participation fee by Institute of complex systems for "Julich Soft Matter Days 2012", Nov 13–16, in Julich, Germany.
- Awarded Silver Medals in Physics, Mathematics and English subjects for best performance in the University level in B.Sc.

Projects Assisted

- Investigation of Interface induced anisotropy in Nanoclay dispersions.
- Phase stability and aging dynamics of Laponite in Ionic liquid.
- Drying induced patters in anisotropic particles.

Publications

1. **Ravi Kumar Pujala**, Nisha Pawar and H. B. Bohidar "Universal sol state behaviour and gelation kinetics in mixed clay dispersions" ***Langmuir* 27, 5193 (2011)**.
2. **Ravi Kumar Pujala**, Nisha Pawar and H. B. Bohidar "Landau theory description of observed isotropic to anisotropic phase transition in mixed clay gels" ***Journal of Chemical Physics* 134, 194904 (2011)**.
3. **Ravi Kumar Pujala**, Nisha Pawar and H. B. Bohidar "Unified scaling behaviour of physical properties of clays in alcohol solutions"***Journal of Colloid and Interface Science* 364, 311 (2011)**.
4. **Ravi Kumar Pujala** and H. B. Bohidar "Ergodicity breaking and aging dynamics in Laponite- Montmorillonite mixed clay dispersions" ***Soft Matter* 8, 6120 (2012)**.
5. **Ravi Kumar Pujala** and H. B. Bohidar "Slow Dynamics, Hydration and Heterogeneity in Laponite Dispersions", ***Soft Matter* 9, 2003 (2013)**.
6. **Ravi Kumar Pujala** and H. B. Bohidar "Kinetics of anisotropic ordering in Laponite dispersions induced by a water–air interface" ***Phys. Rev. E* 88, 052310 (2013)**.
7. **Ravi Kumar Pujala**, Nidhi Joshi and H. B. Bohidar "Spontaneous evolution of self- assembled phases from anisotropic colloidal dispersions", *(under review)*. (*arXiv:1306.1703*)
8. **Ravi Kumar Pujala** and H. B. Bohidar "Revisited phase diagram of high aspect ratio Montmorillonite aging dispersions", *to be submitted*.
9. * **Ravi Kumar Pujala** and H. B. Bohidar "Aging, Slow Dynamics and Anisotropic ordering in Nanoclay Dispersions: A Review" *in preparation*.
10. * Nidhi Joshi, **Ravi Kumar Pujala,** H. B. Bohidar, "Ionic liquid driven phase separation in Nanoclay dispersions and the phase diagram ", *under review*.

    * (not part of the thesis)